Springer Tracts in Modern Physics 109

W0111551

Editor: G. Höhler
Associate Editor: E. A. Niekisch

Editorial Board: S. Flügge H. Haken J. Hamilton
H. Lehmann W. Paul

Springer Tracts in Modern Physics

83 **Pion-Electroproduction.** Electroproduction at Low Energy and Hadron Form Factors
By E. Amaldi, S. P. Fubini, G. Furlan

84 **Collective Ion Acceleration** With contributions by C. L. Olson, U. Schumacher

85 **Solid Surface Physics** With contributions by J. Hölzl, F. K. Schulte, H. Wagner

86 **Electron-Positron Interactions** By B. H. Wiik, G. Wolf

87 **Point Defects in Metals II:** Dynamical Properties and Diffusion Controlled Reactions
With contributions by P. H. Dederichs, K. Schroeder, R. Zeller

88 **Excitation of Plasmons and Interband Transitions by Electrons** By H. Raether

89 Giant Resonance Phenomena in **Intermediate-Energy Nuclear Reactions**
By F. Cannata, H. Überall

90* **Jets of Hadrons** By W. Hofmann

91 **Structural Studies of Surfaces**
With contributions by K. Heinz, K. Müller, T. Engel, and K. H. Rieder

92 **Single-Particle Rotations in Molecular Crystals** By W. Press

93 **Coherent Inelastic Neutron Scattering in Lattice Dynamics** By B. Dorner

94 **Exciton Dynamics in Molecular Crystals and Aggregates** With contributions by
V. M. Kenkre and P. Reineker

95 **Projection Operator Techniques in Nonequilibrium Statistical Mechanics**
By H. Grabert

96 **Hyperfine Structure in 4d- and 5d-Shell Atoms** By S. Büttgenbach

97 **Elements of Flow and Diffusion Processes in Separation Nozzles** By W. Ehrfeld

98 **Narrow-Gap Semiconductors** With contributions by R. Dornhaus, G. Nimtz, and
B. Schlicht

99 **Dynamical Properties of IV–VI Compounds** With contributions by H. Bilz,
A. Bussmann-Holder, W. Jantsch, and P. Vogl

100* **Quarks and Nuclear Forces** Edited by D. C. Fries and B. Zeitnitz

101 **Neutron Scattering and Muon Spin Rotation** With contributions by R. E. Lechner,
D. Richter, and C. Riekel

102 **Theory of Jets in Electron-Positron Annihilation** By G. Kramer

103 **Rare Gas Solids** With contributions by H. Coufal, E. Lüscher, H. Micklitz, and
R. E. Norberg

104 **Surface Enhanced Raman Vibrational Studies at Solid/Gas Interfaces** By I. Pockrand

105 **Two-Photon Physics at e^+e^- Storage Rings** By H. Kolanoski

106 **Polarized Electrons at Surfaces** By J. Kirschner

107 **Electronic Excitations in Condensed Rare Gases**
By N. Schwentner, E.-E. Koch, and J. Jortner

108 **Particles and Detectors** Festschrift for Jack Steinberger
Edited by K. Kleinknecht and T. D. Lee

109 **Metal Optics Near the Plasma Frequency**
By F. Forstmann and R. R. Gerhardts

110* **Electrodynamics of the Semiconductor Band Edge**
By A. Stahl and I. Balslev

---

* denotes a volume which contains a Classified Index starting from Volume 36

F. Forstmann   R. R. Gerhardts

Metal Optics Near the Plasma Frequency

With 30 Figures

Springer-Verlag
Berlin Heidelberg GmbH

Professor Dr. Frank Forstmann
Professor Dr. Rolf R. Gerhardts*

Freie Universität Berlin, Fachbereich Physik
Institut für Theorie der kondensierten Materie, Arnimallee 14
D-1000 Berlin 33, Germany

* *Present address:* Max-Planck-Institut für Festkörperforschung
Heisenbergstraße 1, D-7000 Stuttgart 80, Fed. Rep. of Germany

---

Manuscripts for publication should be addressed to:
Gerhard Höhler

Institut für Theoretische Kernphysik der Universität Karlsruhe
Postfach 6380, D-7500 Karlsruhe 1, Fed. Rep. of Germany

*Proofs and all correspondence concerning papers in the process of publication
should be addressed to:*

Ernst A. Niekisch

Haubourdinstrasse 6, D-5170 Jülich 1, Fed. Rep. of Germany

ISBN 978-3-662-15195-2   ISBN 978-3-540-47175-2 (eBook)
DOI 10.1007/978-3-540-47175-2

Library of Congress Cataloging-in-Publication Data. Forstmann, F. (Frank), 1937–. Metal optics near the plasma frequency. (Springer tracts in modern physics; 109) Includes index. 1. Metals – Surfaces – Optical properties. 2. Plasma frequencies. I. Gerhardts, R. (Rolf), 1949–. II. Title. III. Series. QC1.S797 vol. 109   530 s [530.4'1] 86-22072 [QC176.8.06]

This work is subject to copyright. All rights are reserved, whether the whole or part of the material is concerned, specifically those of translation, reprinting, reuse of illustrations, broadcasting, reproduction by photocopying machine or similar means, and storage in data banks. Under § 54 of the German Copyright Law where copies are made for other than private use, a fee is payable to "Verwertungsgesellschaft Wort", Munich.

© Springer-Verlag Berlin Heidelberg 1986
Originally published by Springer-Verlag Berlin Heidelberg New York in 1986
Softcover reprint of the hardcover 1st edition 1986

The use of registered names, trademarks, etc. in this publication does not imply, even in the absence of a specific statement, that such names are exempt from the relevant protective laws and regulations and therefore free for general use.

2153/3150-543210

Preface

Standard textbooks on optics neglect the fact that light interacting with metal surfaces or interfaces can excite longitudinal plasma waves inside the metal. This phenomenon becomes more important at frequencies of the order of metal plasma frequencies ($\hbar\omega \approx 10$ eV), the region in which the advent of synchrotron radiation has opened up a rapidly growing field of optical experiments. In this volume we demonstrate (Chap. 3) that reflection from and transmission through metal layers, electroreflectance and ellipsometry from metal surfaces, the surface plasmon and eigenmodes in thin layers and spheres, and also the fields near the surface as encountered in photoemission cannot be understood even qualitatively without taking plasma waves into account.

A major aim of this book is to demonstrate that it is not only necessary to include plasma waves in the calculation of the optical response of a variety of systems containing metal surfaces and layers, but that it is also possible to do so within a rather simple, straightforward extension of classical Fresnel optics, the so-called "hydrodynamic approximation" (HD). This HD has a clear physical concept and is easy to handle mathematically. In addition to the familiar transverse electromagnetic waves, it includes longitudinal waves which are also solutions of Maxwell's equations if the wave-number dependence of the dielectric function, i.e. "spatial dispersion", is taken into account in a Drude-type theory. From a theoretical point of view, the HD is the simplest approximate version of "nonlocal optics". To discuss its relations to other phenomenological as well as the more general microscopic approaches is the second major aim of this article. It turns out that for many purposes the HD provides a very useful interpretation scheme, which allows a qualitative understanding of the underlying physics without requiring a complicated theoretical and numerical machinery.

We therefore want to encourage experimentalists to use the HD for the interpretation of experiments and we start with the discussion of this slight extension of standard optics, which has all the features of a "textbook metal optics". We present the method and demonstrate its successful application in many circumstances. This scheme of calculating the optical response of systems with conduction electrons is conceptually no more difficult than calculation of the reflec-

V

tivity of a glass plate and can well be included in graduate lectures on electrodynamics or optics.

The second half of this book sets out several more sophisticated treatments of
the response of a single metal surface to electromagnetic radiation. The relations
between general forms of the response functions and additional boundary conditions
are discussed for a variety of phenomenological models, the HD, the specular reflection model, the "dielectric approximation", the "semi-classical infinite barrier" model and several others. Also, microscopic surface response calculations
based on a quantum mechanical model containing a surface potential for the electrons are presented and their relationships to the HD and the SCIB model are il-
lustrated. Finally, we deal with two surface response functions $d_\perp(\omega)$ and $d_\parallel(\omega)$,
which contain the integrated effect of spatial dispersion and surface properties on
measurable quantities such as reflection coefficient or surface plasmon dispersion.
Their relation to measurements as well as to microscopic or phenomenological response calculations is discussed.

While the second half (Chaps. 4 and 5) of this volume addresses more the reader
who is interested in the theoretical aspects of the metal surface response, the
first three chapters are meant for anybody who comes across optical problems involving conducting surfaces. There should also be a field of technical applications: special optical properties of layered structures involving conductors, optimal
or selective mirrors in the synchrotron radiation frequency range and related problems. Here a successful method for calculating the optical properties of conducting
systems is presented that is simple and transparent enough to become a standard
tool in metal optics.

This book is a result of several years work and would not have been possible in
this form without interaction with other scientists. We gratefully remember and
acknowledge discussions with many of our colleagues in this field, who helped us
to clarify the problems and to reach the point of view from which we treat metal
optics in this article. We would like to mention especially P. Apell, A. Bagchi,
R. Barrera, A.D. Boardman, A.M. Brodsky, A. Eguiluz, P.J. Feibelman, R. Fuchs,
P. Gies, K. Kempa, K.L. Kliewer, D.M. Kolb, A.S. Kondratjev, A.E. Kuchma, S. Liu,
T. Lopez-Rios, S. Lundqvist, T. Maniv, D.L. Mills, G. Mukhopadhyay, H. Raether,
R. Ritchie, W.L. Schaich, H. Stenschke, K. Sturm.

Finally, we acknowledge financial support from the Deutsche Forschungsgemein-
schaft through Sonderforschungsbereich 6.

Berlin · Stuttgart, May 1986   *F. Forstmann · R.R. Gerhardts*

Contents

1. Introduction 1

2. Metal Optics in the Hydrodynamic Approximation 6
   2.1 Plasma Waves, Nonlocality, Spatial Dispersion 6
   2.2 The Hydrodynamic Model for Simple Metals 8
   2.3 The Additional Boundary Conditions 10
   2.4 The Additional Boundary Conditions at a Free Surface 12
   2.5 The Energy Theorem for Combined Transverse and Longitudinal Fields 14
   2.6 The Additional Boundary Conditions at an Interface Between Two Metals
       of Different Electron Density 17
   2.7 Extension to Not-Nearly-Free Electron Metals 19

3. Applications of Nonlocal Metal Optics 22
   3.1 Reflection and Transmission 22
   3.2 Resonances in Thin Metal Films 28
   3.3 The Surface Plasmon Dispersion 29
   3.4 Standing Wave Eigenmodes in Thin Surface Layers 36
   3.5 Optical Properties of Metal Layers on Metal Substrates 41
   3.6 Electroreflectance Spectra at Silver Surfaces 43
   3.7 Ellipsometry from Metal Surfaces 46
   3.8 Resonances in Small Metal Spheres 48
   3.9 The Electric Fields Near the Surface Inside the Metal 52
   3.10 The Photoemission Yield from Metal Surfaces 55
   3.11 Different Additional Boundary Conditions 58

4. Theoretical Concepts and Models of Metal Surface Response 62
   4.1 Additional Boundary Conditions or Susceptibility 62
   4.2 Green's Functions for an Extended Hydrodynamic Model 65
   4.3 The Specular Reflection Model 70
   4.4 Microscopic Response Theory 73
   4.5 Collective and Single-Particle Response in the SCIB Model 79

5. **Description of Nonlocal Effects by the Surface Response Functions**
   $d_\perp(\omega)$ and $d_\parallel(\omega)$ 89
   5.1 Economical Presentation of Experimental Results: $d_\perp(\omega)$, $d_\parallel(\omega)$ 89
   5.2 Boundary Conditions for the Asymptotic Fields 91
   5.3 The Long-Wavelength Limit 96
   5.4 Local Model for the Surface Region 102
   5.5 Nonlocal Layer Model in the Hydrodynamic Approximation 107
       5.5.1 No Surface Layer 110
       5.5.2 Local Limits 112
       5.5.3 Long Wavelength Limit 113
   5.6 Surface Plasmons 116
       5.6.1 Feibelman's Treatment 116
       5.6.2 Additional Surface Plasmon Modes 118
       5.6.3 Experimental Evidence for Multipole Surface Plasmons 120
   5.7 Résumé 123

References 125

Subject Index 131

1. Introduction

In recent years optical methods such as differential reflection spectroscopy and ellipsometry, electroreflection, excitation of surface plasmons, ultraviolet photo-emission, etc., have extensively been applied to characterize clean and adsorbate-covered metal surfaces both in vacuum and in contact with an electrolyte. In the near future, optical methods at ultraviolet frequencies probably will become even more important, since synchrotron radiation as a convenient high intensity radiation is becoming available in many places. Especially for the investigation of metal/electrolyte interfaces optical methods are indispensable, since the power-ful particle beam methods of surface physics require vacuum conditions and are not applicable at a liquid interface.

The optical surface effects we discuss in this article occur in principle for all frequencies of the incident light and are closely related to the perturbation of the surface charge distribution, which is always induced by p-polarized light (i.e. light with a finite normal component of the electric field vector) at the interface between media of different bulk dielectric constants. At a metal surface, these induced screening charges are confined to a narrow but finite surface region, typically a few Angstroms thick or less. In principle, a modification of the charge distribution at the surface, for instance by ad-atoms or an applied static elec-tric field, affects the distribution of optically induced charges and thus the surface electromagnetic fields and the response properties. The optical techniques mentioned in the beginning are so sensitive, that already minute modifications of the surface constitution, for instance by much less than a monolayer of ad-atoms, can be clearly detected. In many cases, for instance for metal electrodes in elec-trolytic cells, these optical techniques provide the most detailed information available.

In this article we will not consider inelastic light scattering which may occur at certain frequencies owing to transitions between particular collective modes or single electron states of the metal or an absorbate.

The aim of optical experiments is to obtain information about the constitution of the surface, e.g. the electronic charge distribution. Clearly, a reliable theoretical basis for the interpretation of such optical surface data is highly desirable. Unfortunately, classical Fresnel optics, which has been applied successfully to many problems of metal optics over many decades, is not sufficient for an adequate understanding of many experimental results on a variety of metal surfaces. In Chap. 3 of this article we discuss several experimental examples which, in the frequency region near the plasma frequency (e.g. $0.5\omega_p \lesssim \omega \lesssim 1.5\omega_p$), clearly refute the predictions of classical optics.

The reason for the failure of classical optics in this context is easy to understand. The standard procedure of classical Fresnel optics /1.1/ is based on the assumption that a boundary between media can be approximated by a plane between two homogeneous regions of different dielectric properties and that the general solutions of Maxwell's equations on both sides of the interface are transverse divergence-free electromagnetic waves, which are matched by the standard boundary conditions. For p polarization, $E_\parallel$ and $D_\perp$, the tangential component of the electric field and the normal component of the displacement field, respectively, must be continuous; therefore the discontinuity of $E_\perp$ and thus the induced surface charge is proportional to the discontinuity of the inverse dielectric constant $1/\varepsilon(\omega)$ across the interface. In order to simulate within this approach the effect of a slightly modified surface, for instance by ad-atoms or an applied electric field, one has to assume a thin surface layer (thickness of the order 1 Å) with a modified dielectric constant $\varepsilon_s(\omega)$. Near the plasma frequency of this surface layer, $1/\varepsilon_s(\omega)$ becomes large and the classical Fresnel theory predicts a large induced charge and a large optical response, which has no correspondence in the experimental findings. In real systems there is no perfect screening and no singular surface charge density. The induced charges are spread over the distance of a screening length, typically of the order of 1 Å in metals, and the induced charges in a thin surface layer are coupled to those in the substrate. As a consequence, at the plasma frequency of the layer material no resonant plasma oscillations are found in a very thin adlayer.

Since the spatial extension of optically induced charges is important for an understanding of difference-spectroscopy experiments near the plasma frequency, a microscopic theory of electromagnetic surface response seems a suitable basis for the interpretation of such experiments. Needless to say, that a full quantum mechanical theory, including lattice and band-structure effects and treating surface and bulk response properties on equal footing, would require a tremendous numerical effort and has not even been attempted. The first ambitious work towards this direction has been presented and recently reviewed by FEIBELMAN /1.2/. He considered free-electron metals within the jellium model, which replaces the metal ions by a homogeneous structureless background of positive charge, thereby neglecting lattice effects, assumed a realistic effective surface potential, which keeps the electrons inside the metal, and calculated the optical response within the random phase approximation (RPA). The results of Feibelman's microscopic theory are in remarkably good agreement with the measured photoyield spectrum of aluminum in a broad frequency interval including the plasma frequency (cf. Sect. 3.10). The microscopic calculation is very involved and dependent on massy numerical work. These complications have prohibited its ready application to a variety of optical problems even for simple metals.

The overwhelming majority of optical surface data is, however, not on free-electron metals but on noble metals and transition metals, where interband transitions are known to be very important for the optical response properties. Then neither the classical Fresnel optics nor the RPA calculations based on the jellium model provide a suitable basis for the evaluation of experimental data: The former cannot describe the spatial spread of induced charges, although the correct bulk dielectric constant can be taken as input. The latter describes the charge distribution near the surface well, but cannot account for the correct bulk response.

The main purpose of this article is to present a detailed discussion of the merits, but also of the limitations, of a simple phenomenological generalization of classical optics, which can describe both, the spatial spread of induced charges and the correct bulk dielectric properties. The idea of this so-called "hydrodynamic approximation" (HD) is to include among the fundamental solutions of Maxwell's equations, which build up the electromagnetic field in the metal, not only the divergence-free transverse waves of Fresnel optics but also the longitudinal wave solutions, which occur if the spatial dispersion, i.e. the k dependence in Fourier-space, of the longitudinal dielectric function $\varepsilon_\ell(k,\omega)$ is taken into account. With the model assumption

$$\varepsilon_\ell(k,\omega) = \varepsilon_b(\omega) - \omega_n^2 / [\omega(\omega + i\gamma) - \beta k^2] \tag{1.1}$$

of the HD and $\varepsilon_t = \varepsilon_\ell(0,\omega)$ for the transverse dielectric function, the longitudinal eigenmodes of Maxwell's equations are plasma waves and the optically induced, spatially spread charge density, the divergence of the longitudinal electric field, is approximated as plasma oscillations of the conduction electrons. The material parameters in (1.1) account for the correct bulk response properties. Just as the classical Fresnel optics, the HD works with matching conditions for the fields at an interface between two media with different, but on both sides spatially constant material parameters. Since the amplitudes of transverse and longitudinal waves must be determined, besides the standard matching conditions "additional boundary conditions" (ABC) are formulated.

In Chap. 2 we explain the hydrodynamic approximation in detail and in Chap. 3 we present a large number of different experimental examples which could be understood and interpreted within this approach, whereas classical Fresnel optics failed. We want to emphasize that this success is not accidental and that the inclusion of plasma waves in metal optics is not just a theoretical trick. On the contrary, there is experimental evidence for the optical excitation of plasma waves in metals and ample justification from microscopic models.

Historically, plasma waves were predicted theoretically by BOHM and PINES /1.3/ in the fifties as eigenmodes in a homogeneous metal. Subsequently they were studied by "longitudinal probes", i.e. by fast electrons transmitted through metal foils, which loose energy of quanta $\hbar\omega_p$ on excitation of plasmons /1.4/. In 1964 SAUTER /1.5/ pointed out that these longitudinal plasma waves should be included in metal optics, since they are solutions of Maxwell's equations if spatial dispersion, i.e. non-locality of the longitudinal electronic response is taken into account. Sauter required as ABC the continuity of the normal component of the current density and published in 1967 a calculation /1.6/ of reflection from a metal surface, together with a microscopic justification given by FORSTMANN /1.7/ using Boltzmann's equation. On the basis of essentially the same macroscopic approach, MELNYK and HARRISON /1.8/ predicted a few months later the possibility of resonant excitation of plasmons in thin metal films by electromagnetic waves. Such resonances were observed experimentally in 1970/71 /1.9 , 10/, and were the first direct proof that plasma waves are indeed excited optically.

Of course, the hydrodynamic approximation is a phenomenological approach and has its limitations. Collective plasma oscillations are only one possible response mode of the electron system. Optical excitation of individual electron-hole pairs is also possible near a surface and provides an additional response and absorption mechanism absent in the free electron model of bulk metals. Such effects are not included automatically in the HD, but they can be simulated by a suitable choice of parameters, e.g. for the damping in a surface layer. In Chap. 4 we discuss the limitations of the HD and we review other phenomenological and microscopic approaches to metal optics. From the comparison of different approaches we get the impression, that the HD is easiest to handle and, eventually with a suitably chosen surface layer, flexible enough to provide a good, at least qualitative, understanding of surface response properties. It is interesting to note that the physical content of Feibelman's microscopic result for the photoyield of aluminum was explained by model calculations within the HD, as will be discussed in Sect. 3.10.

In the final Chap. 5 we study the surface response functions $d_\perp(\omega)$ and $d_\parallel(\omega)$, which were introduced in this form by FEIBELMAN /1.2/ and are now considered by many authors /1.11/. These surface response functions contain the full information about the nonlocal surface response and determine the macroscopic quantities, which can be measured in different experimental situations. They provide a useful frame for the presentation of experimental results, without imposing restrictive model assumptions, and are convenient for the comparison of experiment and theory. To clarify the physical implications of the peculiar frequency dependence of the sur-

4

face response function $d_\perp(\omega)$ /1.2/, we evaluate $d_\perp$ in the HD and discuss the result in the context of "multipole surface plasmons" (cf. Sects. 5.5 , 6). We want to notice that we mean eigenmodes of the full Maxwell equations in the system, including the nonlocal response of the metal, if we talk about "plasmons" or "plasma waves". We avoid the phrases "plasmon polaritons", "surface plasmon polaritons", etc. used by some authors, mainly in the context of elementary excitations in semiconductors but also in metal optics.

Several review articles covering different aspects of electromagnetic surface response properties of metals have been published already and will be mentioned in the following chapters. In this article we tried to keep mere repetition of previous reviewing work at a minimum and to concentrate on aspects, which in previous reviews were either not treated at all, or from a different point of view.

2. Metal Optics in the Hydrodynamic Approximation

At higher frequencies inertia prevents the electrons from instantaneous screening; the assumption of standard optics that the metal is free of charges is not valid any more. The metal can sustain *plasma waves*, charge density waves with longitudinal electric fields. These waves are homogeneous solutions of Maxwell's equations and should therefore be included in the general solution and consequently be considered in optics. This reasoning was put forward by SAUTER /2.1 , 2/ and he suggested, that a minor extension of standard optics can include the plasma waves. The *hydrodynamic approximation* (HD) provides this extension. The interface is approximated by a plane between homogeneous regions, the general solutions in the homogeneous regions are matched by boundary conditions. In view of its simplicity and its success (Chap. 3) this method can be considered the textbook metal optics.

2.1 Plasma Waves, Nonlocality, Spatial Dispersion

Maxwell's equations for metals will be used in the form:

$$\text{curl } \mathbf{B} = \frac{1}{c}\frac{\partial \mathbf{E}}{\partial t} + \frac{4\pi}{c}\,\mathbf{j} \qquad (2.1) \qquad\qquad \text{div } \mathbf{E} = 4\pi\rho \qquad (2.3)$$

$$\text{curl } \mathbf{E} = -\frac{1}{c}\frac{\partial \mathbf{B}}{\partial t} \qquad (2.2) \qquad\qquad \text{div } \mathbf{B} = 0 \qquad (2.4)$$

$$\text{curl curl } \mathbf{E} = \text{grad (div } \mathbf{E}) - \Delta\mathbf{E} = -\frac{1}{c^2}\frac{\partial^2 \mathbf{E}}{\partial t^2} - \frac{4\pi}{c^2}\frac{\partial \mathbf{j}}{\partial t} \quad . \qquad (2.5)$$

We will deal with nonmagnetic metals, $\mu = 1$, $\mathbf{B} = \mathbf{H}$. All current densities $\mathbf{j}$ and charge densities ρ appear explicitly in (2.1 , 3). Usually we imagine the metal electrons as free conduction electrons; only in the application to silver (Sects. 2.7, 3.4 - 7) we generalize by including polarization currents and densities due to bound electrons.

Maxwell's equations describe the fields due to known currents and charges. They have to be completed by material equations stating which currents and charges are produced by given fields. We consider systems which respond linearly to the perturbing field. The material equations can be written in the form:

$$j(r,t) = \int \overset{\leftrightarrow}{\sigma}(r,r',t-t')\ E(r',t')\ d^3r'\ dt' \tag{2.6}$$

$$D(r,t) = \int \overset{\leftrightarrow}{\varepsilon}(r,r',t-t')\ E(r',t')\ d^3r'\ dt$$
$$= E(r,t) + 4\pi \int \overset{\leftrightarrow}{\chi}(r,r',t-t')\ E(r',t')\ d^3r'\ dt' \tag{2.7}$$

where E is the selfconsistent electric field appearing in Maxwell's equations. The conductivity σ, the dielectric function ε or the polarizability χ are related by linear response theory /2.3/ to a microscopic description of the system. The response functions are generally tensors and 'nonlocal' in space and time, i.e. the current at r and t is determined by the field E in a certain neighbourhood of r (in principle by the field distribution in the whole space) and by the field distributions at earlier times. For fields varying over distances large compared to the lattice spacing the bulk metal can be taken as a translationally invariant system. Then σ and ε depend only on the difference $r-r'$, (2.6,7) are convolutions and have a product form after Fourier transformation:

$$j(k,\omega) = \overset{\leftrightarrow}{\sigma}(k,\omega)\ E(k,\omega) \tag{2.8}$$

$$D(k,\omega) = \overset{\leftrightarrow}{\varepsilon}(k,\omega)\ E(k,\omega) = [1 + 4\pi\overset{\leftrightarrow}{\chi}(k,\omega)]\ E(k,\omega) \tag{2.9}$$

$$D = E - \frac{4\pi}{i\omega}\ j = \overset{\leftrightarrow}{\varepsilon}E\ ,\qquad \overset{\leftrightarrow}{\varepsilon} = 1 - \frac{4\pi}{i\omega}\overset{\leftrightarrow}{\sigma} = 1 + 4\pi\overset{\leftrightarrow}{\chi}\ . \tag{2.10}$$

A system is called *spatially dispersive* /2.4/, if its response functions depend on the wave vector k which is mathematically related to the nonlocality in (2.6,7).

Much effort has been spent on calculating $\sigma(k,\omega)$ or $\varepsilon(k,\omega)$ for bulk metals /2.5/. But the knowledge of the bulk response is not sufficient to solve the problem of metal optics, since the surface breaks the translational invariance of the system. Then σ and ε depend on r and r' separately and not only on the difference $r-r'$. There have been many attempts to derive the response near the surface from the known response function of the bulk. These studies are set in relation in Chap. 4. The HD applied by SAUTER /2.1,2/ and FORSTMANN and STENSCHKE /2.6/ is the simplest approach and the one used most successfully.

The approximation to (2.6) used in standard optics neglects nonlocality all together, takes the local proportionality $j(r) = \sigma(\omega)E(r,\omega)$ and assumes σ to have the constant bulk value up to the surface plane. Maxwell's equations are then solved by transversal waves in the two homogeneous regions (e.g. inside and outside the metal). For p-polarized light the two free amplitudes of the outgoing waves in the general solutions are determined from two boundary conditions:

tangential component of **E** continuous $\qquad$ (2.11)

normal component of $\mathbf{D} = \mathbf{E} - \frac{4\pi}{i\omega}\mathbf{j}$ continuous . $\qquad$ (2.12)

These boundary conditions are derived from integrating (2.2) and div **D** = 0 from
(2.1). Condition (2.12) is equivalent to the continuity of the tangential component
of **H**, which has to be used when $D_n = 0$ in the case of s polarization. The result
of this procedure are the Fresnel reflection and transmission amplitudes (see /2.7 , 8/
and Sect. 3.1).

The Fourier transform of (2.5) is

$$-\mathbf{k}(\mathbf{k}\cdot\mathbf{E}) + k^2\mathbf{E} = \frac{\omega^2}{c^2}\varepsilon(\mathbf{k},\omega)\,\mathbf{E} \quad . \qquad (2.13)$$

Transverse waves, $\mathbf{k}\cdot\mathbf{E} = 0$, solve this equation for $k^2 = \varepsilon\omega^2/c^2$, while longitudinal
waves, $\mathbf{k}\,\|\,\mathbf{E}$, make the left hand side zero and nontrivial solutions require $\varepsilon(\mathbf{k},\omega)$
= 0. In metals these longitudinal waves are called plasma waves. They are homoge-
neous solutions of Maxwell's equations and are a consequence of the k dependence of
ε, of spatial dispersion, of nonlocal response functions. Therefore inclusion of
plasma waves in optics mean taking into account in some approximation the nonlocality of the response.

2.2 The Hydrodynamic Model for Simple Metals

When ε depends on the wave vector **k**, an interface which breaks the translational
invariance causes additional problems. Electrodynamics near an interface must be
formulated in **r** space, but the Fourier transform from $\varepsilon(\mathbf{k},\omega)$ to $\varepsilon(\mathbf{r}-\mathbf{r}',\omega)$ is only
simple with translational invariance. Near a surface ε depends on **r** and **r**' separate-
ly, not only on the difference $\mathbf{r}-\mathbf{r}'$, and the proper dependence cannot mathematically be derived from a knowledge of the bulk $\varepsilon(\mathbf{k})$. The new physics near the interface has to be explicitly introduced (Sects. 2.3 , 4). Several ways to treat a surface
of a spatially dispersive metal are discussed in the following chapters. The hydrodynamic approximation (HD) is the simplest which proved to be successful.

The essential steps of the HD consist of approximating the k dependence of the
bulk ε by the first term proportional to k^2 and formulating additional boundary conditions at an interface. The k^2 term is the first in a small k (long wave length)
expansion for a system with inversion symmetry. In order to allow for the existence
of plasmons it suffices to keep only this k^2 term in the response function for *lon-
gitudinal* perturbations: $\varepsilon_\|(k,\omega)$, $\sigma_\|(k,\omega)$, while the k dependence is dropped com-
pletely in the transverse dielectric function $\varepsilon_\perp(\omega)$. This further approximation is

usually taken for metals, while for semiconductors the discussion of "polaritons" includes also "transverse polaritons" due to the k dependence of $\varepsilon_\perp$ /2.4 , 9 , 10/.

The necessary transition to **r** space is best accomplished by recognizing that the approximated $\sigma_\parallel(k,\omega)$, $\sigma_\perp(\omega)$ are the response functions, the Green functions, of the hydrodynamic equation of motion of the metal electrons, of the *hydrodynamic model* which was introduced by BLOCH /2.11/. As discussed by PINES and NOZIERES /2.12/ the response of a homogeneous electron gas to time dependent density perturbations is analogous to that of a fluid in two frequency regions: The many collision regime $\omega\tau \ll 1$ (τ the collision time) where the electrons scatter frequently during one period and relax to local equilibrium, and the collisionless regime $\omega\tau \gg 1$ where the electron gas reacts like a jelly due to the strong coupling by the long range Coulomb forces. In these frequency regimes, the electron motion can be approximately calculated from

$$\frac{d}{dt}(n_0 m \mathbf{v}) = n_0 e \mathbf{E} - m\cdot\gamma\mathbf{v} - \text{grad } p \tag{2.14}$$

with n_0 the equilibrium electron density, e the elementary charge, m the electron mass, $\gamma = 1/\tau$ determining a friction like damping of the motion and p being the electron gas pressure. grad p is proportional to the gradient of the electron density. Multiplication by e/m yields the equation for the macroscopic charge and current densities ρ and **j**:

$$\frac{d}{dt}\mathbf{j} = \frac{\omega_p^2}{4\pi}\mathbf{E} - \gamma\mathbf{j} - \beta\,\text{grad }\rho \tag{2.15}$$

with $\omega_p^2 = 4\pi n_0 e^2/m$. $\tag{2.16}$

Equation (2.15) is the material equation which together with the field equation (2.5) forms a closed set of differential equations. It is an extended Drude model, extended by the pressure term on the r.h. side which causes the propagation of plasma waves, which takes account of nonlocality and spatial dispersion. The forces, that drive the current, are the electric field **E**, a friction force and the Fermi gas pressure. In comparison to standard optics the new plasma wave eigenmode is due to the new force grad p in the equation of motion. The damping constant γ describes all energy losses from the system of fields and collective electron motion (see Sect. 2.5).

The hydrodynamic model (2.15) can be derived, for instance, from a Boltzmann equation for the electron gas /2.13/. The relation to microscopic electron gas theories can be made even closer by realizing, that perturbations $\propto \exp(i\mathbf{kr} - \omega t)$ yield from (2.15) with div **j** = $- \partial\rho/\partial t$:

for transverse **E**: $$\mathbf{j}_\perp = \frac{i\omega_p^2}{4\pi(\omega + i\gamma)}\mathbf{E}_\perp = \sigma_\perp\mathbf{E}_\perp \tag{2.17}$$

for longitudinal E:
$$j_\parallel = \frac{i\omega \omega_p^2}{4\pi[\omega(\omega + i\gamma) - \beta k^2]} E_\parallel = \sigma_\parallel E_\parallel \quad . \tag{2.18}$$

Together with (2.10) we get

$$\epsilon_\perp(\omega) = 1 - \frac{\omega_p^2}{\omega(\omega + i\gamma)} \quad ; \qquad \epsilon_\parallel(k,\omega) = 1 - \frac{\omega_p^2}{\omega(\omega + i\gamma) - \beta k^2} \quad . \tag{2.19}$$

The response of the model (2.15) is the same as that of any sophisticated electron gas model in the small k approximation. The three model parameters ω_p, γ and β can be read off from such an expansion. Because the damping of the collective motion due to incoherent electron hole excitations, phonons, impurities, surface roughness etc. is never treated fully in a microscopic theory, γ is better taken from experiments, e.g., from the width of energy loss spectra /2.5/. The low frequency value of γ taken from the d.c. conductivity is usually too small by an order of magnitude. As discussed already by BLOCH /2.14/ the factor β in the last term of (2.15) is different for low frequencies than for high frequencies. For a free electron model $\beta = v_F^2/3$ for low frequencies and $\beta = 3v_F^2/5$ for high frequencies of the order of ω_p. The pressure for free electrons is only due to the kinetic energy, the Fermi energy. For high frequencies the electrons do not relax to the lowest states around E_F. Therefore the total energy increases more rapidly with increasing n causing a higher pressure. A derivation of the high frequency factor from the Boltzmann equation can be found in /2.13/ (see also /2.15/).

Setting $\epsilon = 0$ in (2.19) and neglecting damping the HD yields the well known bulk plasmon dispersion /2.16/:

$$\omega^2 = \omega_p^2 + \beta k^2 \quad . \tag{2.20}$$

The relation of the HD to more extended theories will be further discussed in Chaps. 4,5. Its application to static surface problems like image potential, van der Waals forces and surface energy has been reviewed by BARTON /2.17/.

2.3 The Additional Boundary Conditions

In Sect. 2.2 the material equations in the HD have been cast into three forms: (a) the differential equation (2.15), (b) the product (2.17), (2.18) in Fourier representation and (c) the integral relation (2.6) which can be considered the inversion of (2.15) by means of a Green function or the inversion of (2.17), (2.18) by Fourier transformation. Form (b) is only valid for a translationally invariant system. To approach the surface we need version (a) or (c) in r space.

The surface region has in principle response properties different from those of the bulk. How to understand and to model the surface response is still a matter of research and debate (see Chap. 4). In optics one has long tried to get away with the assumption, that a surface or interface is a plane separating two homogeneous regions. We extend this model to spatially dispersive systems. Only for the material equation (2.15) it is obvious that such a model simply requires constancy of the parameters, ω_p, γ, β in each region up to the interface. The "nearly local" formulation of nonlocality by spatial derivatives can be carried from the bulk right up to the surface. In order to reach the Green function formulation (2.6) one needs boundary conditions for the current at the interface. These *additional boundary conditions* (ABC) are statements about the surface response, a surface model in its most reduced form. They come into play in a more lucid way, when the general solutions of Maxwell's equations together with the differential material equation (2.15) is matched across the interface of parameter discontinuity.

The dispute in the past around the ABCs /2.18 - 24/ can probably be settled in the following way: If the surface model is formulated on the level of Schrödinger's equation with a surface potential, one can derive $\sigma(\mathbf{r},\mathbf{r}',\omega)$ needed in (2.6) and no ABCs are necessary. If on the other hand, one starts from any relation between field and current in the bulk, further model assumptions for a surface have to be introduced. If the material parameters determining the bulk response shall be taken unchanged up to the interface (homogeneous regions) ABCs are the explicit way to define the surface model.

To calculate the response of a metal surface from microscopic quantum mechanical models is a formidable task of heavy computation. It is therefore desirable to derive the answer to questions of metal optics from macroscopic relations between fields, current and charge densities. Such a phenomenological approach is simpler to handle, often more transparent and easier adaptable to real experiments. The HD with ABCs is the most successful approximation for this purpose. The ABCs cannot be derived from first principles but only from heuristic arguments. Their usefulness has to be proven by comparison to experiments which are sensitive to effects of spatial dispersion, or to microscopic calculations. They are useful, if they can predict the right amplitude and phases of reflected and transmitted waves, i.e. predict the fields far away from the interface. In principle it is not clear, that simple boundary conditions, which do not depend on frequency and angle, can provide these answers at all for a variety of experimental conditions. We introduce ABCs in Sects. 2.4 , 6 and demonstrate in Chap. 3 their application in several circumstances. It turns out that even the fields in the neighbourhood of the surface can be derived successfully from the HD (Sect. 3.9). The ABCs used in Chaps. 2 , 3 are those most widely tested. In Sect. 3.11 we discuss an objection put forward by BOARDMAN and RUPPIN /2.24/.

2.4 The Additional Boundary Conditions at a Free Surface

Within the hydrodynamic approximation a homogeneous region of a free electron metal is described by

$$\nabla(\nabla \cdot \mathbf{E}) - \nabla^2 \mathbf{E} = -\frac{1}{c^2}\frac{\partial^2 \mathbf{E}}{\partial t^2} - \frac{4\pi}{c^2}\frac{\partial \mathbf{j}}{\partial t} \tag{2.5}$$

$$\frac{\partial \mathbf{j}}{\partial t} = \frac{\omega_p^2}{4\pi}\mathbf{E} - \gamma\mathbf{j} - \frac{\beta}{4\pi}\nabla(\nabla \cdot \mathbf{E}) \qquad (\beta = \frac{3}{5}v_F^2) \quad . \tag{2.15}$$

Under the conditions of optics, which usually specify the plane of incidence (x, z plane), the polarization (x, z plane for p-polarized light), the frequency ω and the angle of incidence or the tangential component k_x of the wave vector $\mathbf{k}$, the general solution of (2.5, 15) is:

$$\mathbf{E}(\mathbf{r},t) = e^{i(k_x x - \omega t)}[\mathbf{E}_1 e^{i\lambda z} + \mathbf{E}_3 e^{-i\lambda z} + \mathbf{E}_2 e^{i\eta z} + \mathbf{E}_4 e^{-i\eta z}] \tag{2.21}$$

$$\mathbf{j}_\perp(\mathbf{r},t) = \frac{i}{4\pi}\frac{\omega_p^2}{\omega + i\gamma}[\mathbf{E}_1 e^{i\lambda z} + \mathbf{E}_3 e^{-i\lambda z}]\, e^{i(k_x x - \omega t)} = \sigma_\perp \mathbf{E}_\perp \tag{2.22}$$

$$\mathbf{j}_\parallel(\mathbf{r},t) = \frac{i\omega}{4\pi}[\mathbf{E}_2 e^{i\eta z} + \mathbf{E}_4 e^{-i\eta z}]\, e^{i(k_x x - \omega t)} = \sigma_\parallel \mathbf{E}_\parallel \tag{2.23}$$

$$\mathbf{j}(\mathbf{r},t) = \mathbf{j}_\perp(\mathbf{r},t) + \mathbf{j}_\parallel(\mathbf{r},t) \tag{2.24}$$

$$\frac{\omega^2}{c^2}\left(1 - \frac{\omega_p^2}{\omega(\omega + i\gamma)}\right) = k_x^2 + \lambda^2 \quad ; \qquad \lambda = \sqrt{\frac{\omega^2}{c^2}\left(1 - \frac{\omega_p^2}{\omega(\omega + i\gamma)}\right) - k_x^2} \tag{2.25}$$

$$\lambda_0 = \sqrt{\frac{\omega^2}{c^2} - k_x^2} \tag{2.26}$$

$$1 - \frac{\omega_p^2}{\omega(\omega + i\gamma) - \beta(k_x^2 + \eta^2)} = 0 \quad ; \qquad \eta = \sqrt{\frac{1}{\beta}[\omega(\omega + i\gamma) - \omega_p^2] - k_x^2} \tag{2.27}$$

$$\mathbf{E}_1 = E_1(-\lambda/k_x, 0, 1) \quad ; \qquad \mathbf{E}_3 = E_3(\lambda/k_x, 0, 1) \tag{2.28}$$

$$\mathbf{E}_2 = E_2(k_x/\eta, 0, 1) \quad ; \qquad \mathbf{E}_4 = E_4(-k_x/\eta, 0, 1) \tag{2.29}$$

$$\sigma_\perp = \frac{i}{4\pi}\frac{\omega_p^2}{\omega + i\gamma} \quad ; \qquad \sigma_\parallel = \frac{i\omega}{4\pi} \quad ; \qquad \varepsilon(\omega) = 1 - \frac{\sigma_\perp}{\sigma_\parallel} = 1 - \frac{\omega_p^2}{\omega(\omega + i\gamma)} \quad . \tag{2.30}$$

We have assumed translational invariance parallel to the x, y plane, i.e. a surface or interface will be a plane z = const. The wave vectors have only x and z components. With the notation (2.28, 29) and (2.21) $\mathbf{E}_1$ and $\mathbf{E}_3$ describe transversal and $\mathbf{E}_2$ and $\mathbf{E}_4$ longitudinal waves. The waves are polarized in the plane of incidence, the x , z plane. This polarization is called p polarization (parallel to the plane of incidence). Only in this case the plasma waves will acquire a nonzero amplitude. For s polarization (senkrecht, **E** only with y component) the whole problem reduces to local standard optics (see Sect. 3.1).

The total electric field is the sum of transversal and longitudinal fields (2.21), each field induces its separate current (2.22 , 23) and the total current is the sum of both (2.24). Formally σ is a tensor with the wave vector as one principal axis. The dispersion formulas (2.25 , 27) lead to complex z components λ and η of the wave vectors when ε is complex (γ $\neq$ 0 in our model). For different frequencies and angles of incidence either the real or the imaginary part of λ and η will dominate. For ω < ω$_p$ the transverse and longitudinal fields essentially decay exponentially inside the metal, while for ω > ω$_p$ the waves can propagate unless k$_x$ is so large, the direction of incidence so flat, that total reflection occurs. We choose λ, λ$_0$ and η with nonnegative imaginary and real parts.

We first discuss a surface of a metal halfspace towards vacuum (Fig. 2.1). Inside the metal **E**$_3$ = **E**$_4$ = 0 in (2.21) because waves are only leaving the surface. Outside we have the indicent wave **E**$_0$ and the reflected wave **E**$_r$. E$_r$, E$_1$ and E$_2$ are the three unknown amplitudes of the general solution. Three boundary conditions are required for their determination.

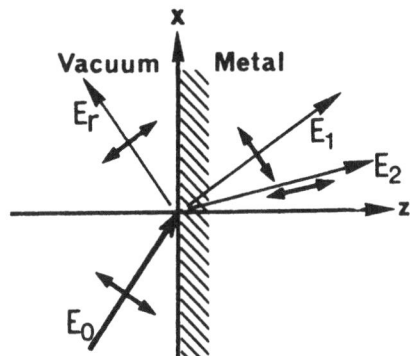

Fig. 2.1. Contributions to the general solution at a free metal surface

Integration of Maxwell's equations across the boundary and the requirement of *finite* fields leads to two independent conditions, the continuity of the tangential component of the electric field and continuity of the normal component of the displacement D$_n$ = E$_n$ - (4π/iω)j$_n$. From these follow the continuity of B$_n$ and of H$_{tg}$ (for fields simply periodic in time and periodic parallel to the interface). In standard optics the proportionality **j** = σ**E** and **D** = ε**E** implies a discontinuity of E$_n$ and j$_n$ following from the continuity of D$_n$. The discontinuity of E$_n$ is equivalent to a singular surface charge density; a discontinuous j$_n$ needs a singular sink or source in the surface plane. These implications of standard optics disagree with any microscopic model of a metal surface, which assumes all charge densities finite. For models with a surface defined by a sharp step of electron density, the normal current goes to zero at that step, while in quantum mechanical models j_n dies out in a surface region (see Chaps. 4, 5). The infinite sink or source of normal current implied by standard optics screens the metal interior from any charge density fluctuations. In reality an oscillating field with a normal component at the surface will induce charge density fluctuations near the surface, which act as the source of the plasma waves. Therefore, in order to couple to the plasma waves properly the singular surface charge should be abandoned. A further argument may be the interpretation of spatial dispersion as caused by the Fermi gas pressure, according to (2.15). A singular surface charge would be either a separate system contradicting the increase of counter forces when charge is accumulated, or would have an infinite pressure.

This kind of arguments led SAUTER /2.1, 2/ to the requirement that there should be no singular surface charge, that j_n should go to zero at a free surface and that E_n should be continuous. These requirements are compatible with the continuity of D_n because σ is a tensor with independent components $\sigma_\perp$ and $\sigma_\parallel$ (2.30), so continuity of E_n and j_n are two independent requirements. For a free metal surface the set of boundary conditions can be written as

Continuity of $E_{tangential}$ (2.31a)

Continuity of E_{normal} (2.31b)

$$j_{normal} = \sigma_\perp E_{\perp n} + \sigma_\parallel E_{\parallel n} = 0 \quad . \tag{2.31c}$$

From (2.31b) and (2.31c) follows that $D_{normal} = E_n - (4\pi/i\omega)j_n$ is continuous.

2.5 The Energy Theorem for Combined Transverse and Longitudinal Fields

From Maxwell's equations (2.1, 2) we obtain the energy theorem

$$-\frac{\partial}{\partial t}\left(\frac{1}{8\pi}E^2 + \frac{1}{8\pi}B^2\right) = \text{div}\left(\frac{c}{4\pi}\mathbf{E} \times \mathbf{B}\right) + \mathbf{E} \cdot \mathbf{j} \quad . \tag{2.32}$$

This equation is usually interpreted by saying that the decrease of the energy density on the left is caused by the energy current leaving the volume element plus the Ohmic loss $\mathbf{E} \cdot \mathbf{j}$ in the volume element. The longitudinal waves have no magnetic field $\mathbf{B}$ according to (2.2). If there is only a longitudinal wave, which is as we saw a solution of Maxwell's equation, is there no energy current? In fact in this case the energy current and additional coherent contributions to the energy density in the electron system, which periodically are transformed back into field energy, are contained in $\mathbf{E} \cdot \mathbf{j}$. This is also apparent if one calculates the field in the example of Fig. 2.1 and evaluates $\mathbf{E} \cdot \mathbf{j}$ for frequencies above ω_p. One gets large oscillations of $\mathbf{E} \cdot \mathbf{j}$ with strongly negative as well as positive values (/2.25/, Sects. 3.9 , 10). This result indicates that $\mathbf{E} \cdot \mathbf{j}$ is not the energy loss density of our system, which must be positive definite. It is not difficult to reformulate $\mathbf{E} \cdot \mathbf{j}$ and (2.32) with the help of (2.15) in a way, that the energy current of the plasma wave, the positive definite energy loss density and additional terms of the energy density are revealed /2.6/. Multiplying (2.15) by $\mathbf{j}$ and using the continuity equation we can rearrange terms to obtain

$$\mathbf{E} \cdot \mathbf{j} = \frac{\partial}{\partial t} \left(\frac{2\pi}{\omega_p^2} j^2 + \frac{2\pi}{\omega_p^2} \beta\rho^2 \right) + \mathrm{div}\left(\frac{4\pi\beta}{\omega_p^2} \mathbf{j}\rho \right) + \frac{4\pi\gamma}{\omega_p^2} j^2 \quad . \tag{2.33}$$

(2.33) combined with (2.32) yields

$$-\frac{\partial}{\partial t} \left(\frac{1}{8\pi} E^2 + \frac{1}{8\pi} B^2 + \frac{2\pi}{\omega_p^2} j^2 + \frac{2\pi}{\omega_p^2} \beta\rho^2 \right) = \mathrm{div}\left(\frac{c}{4\pi} \mathbf{E} \times \mathbf{B} + \frac{4\pi\beta}{\omega_p^2} \mathbf{j}\rho \right) + \frac{4\pi\gamma}{\omega_p^2} j^2 \quad . \tag{2.34}$$

That this arrangement of terms is the proper one can be seen from interpretation of the new contributions:

$$E_{kin} = \frac{2\pi}{\omega_p^2} j^2 = \frac{m}{2n_0 e^2} (n_0 e v)^2 = n_0 \frac{m}{2} v^2 \tag{2.35}$$

is the kinetic energy density due to the drift velocity of the current carrying electrons.

$$E_{pot} = \frac{2\pi}{\omega_p^2} \beta\rho^2 = \frac{m}{2n_0} \beta(n - n_0)^2 \tag{2.36}$$

is the potential energy density of a region with perturbed density n in direct analogy to a sound wave field /2.26/. This potential energy density can be found by integrating the work against the pressure force in the equation of motion (2.15). If in a homogeneous continuum of charge density n_0 we introduce a field of displacements $\mathbf{s}$ ($\mathbf{s} \neq 0$ only in a finite region), the work against the electron gas pressure is

$$\delta \int E_{pot} \, d^3r = \int -\mathbf{K} \cdot \delta\mathbf{s} \, d^3r = \int m\beta(\mathrm{grad}\ n \cdot \delta\mathbf{s}) \, d^3r$$

$$= m\beta \int \mathrm{div}[(n - n_0)\delta\mathbf{s}] \, d^3r - \int m\beta(n - n_0) \, \mathrm{div}\ \delta\mathbf{s} \, d^3r \quad . \tag{2.37}$$

The first integral vanishes when carried out over a large surface where $n - n_0 = 0$, further is $-n_0 \, \mathrm{div}\ \delta\mathbf{s} = \delta n = (n - n_0)$.

$$\delta \int E_{pot} \, d^3r = \int m\beta(n - n_0) \frac{1}{n_0} \delta(n - n_0) \, d^3r = \delta \int \frac{m}{2n_0} \beta(n - n_0)^2 \, d^3r \qquad (2.38)$$

which justifies the interpretation (2.36). The additional energy current density can as well be cast into a form, which shows its analogy to the sound wave case /2.26/:

$$Q = \frac{4\pi\beta}{\omega_p^2} \mathbf{j}\rho = \beta m \mathbf{v}\delta n = \mathbf{v}\delta p \quad . \qquad (2.39)$$

The last equality stems from the electron gas pressure term in (2.15) which is responsible for spatial dispersion:

$$\text{grad } p = \frac{m}{e} \beta \text{ grad } \rho = m\beta \text{ grad } n \qquad (2.40)$$

$$\frac{\partial p}{\partial n} = m\beta \quad . \qquad (2.41)$$

The last term in (2.34) is the energy loss density of our system due to the damping term in the equation of motion:

$$\text{Energy absorption density} = \frac{4\pi\gamma}{\omega_p^2} \mathbf{j}^2 \quad . \qquad (2.42)$$

Here all effects contribute which destroy the coherent hydrodynamic motion of the electrons: phonons and defects as well as incoherent single electron excitations, e.g. photoemitted electrons (see Sect. 3.10). The friction term γ opens a sink of energy to which energy is lost. All other energies are treated explicitly by our system of equations, especially the coherent electron excitations, which form the plasma wave. Therefore (2.32) in conjunction with (2.33) can be read as follows: The decrease of field energy density is due to a flow of field energy out of the volume (Poynting's vector) and to the interaction of the electric field with the electrons. This interaction increases the energy density of the electron system in a coherent manner, causes a longitudinal energy flow carried away with the electrical current, and only the last term in (2.33) is lost for ever from the coupled system of electromagnetic fields and collectively moving charges. All the other energies are transformed back and forth between the fields and the electron system during one period of oscillation.

In several papers studying metal surfaces, the term $\mathbf{E} \cdot \mathbf{j}$ or $\frac{1}{2}(\mathbf{E}^* \cdot \mathbf{j} + \mathbf{E} \cdot \mathbf{j}^*)$ = Re$\{\mathbf{E}^* \cdot \mathbf{j}\}$ for the time average has been used as "absorption density". It has been argued /2.27/ that because the integral of Re$\{\mathbf{E}^* \cdot \mathbf{j}\}$ over the whole metal yields exactly the energy which is missing on reflection, Re$\{\mathbf{E}^* \cdot \mathbf{j}\}$ must be the absorption density. But this argument is not conclusive. Taking Re$\{\mathbf{E}^* \cdot \mathbf{j}\}$ from (2.33) the purely imaginary time derivative in (2.33) does not contribute and we get

$$\int\limits_V \text{Re}\{\mathbf{E}^* \cdot \mathbf{j}\}\, d^3r = \int\limits_V \frac{4\pi\gamma}{\omega_p^2}\, |\mathbf{j}|^2\, d^3r + \int\limits_V \text{div}\!\left(\frac{4\pi\gamma}{\omega_p^2}\, \mathbf{j}\rho\right) d^3r \quad . \tag{2.43}$$

The last term vanishes when the integral is transformed into a surface integral and evaluated where $j_n = 0$. Therefore the integral on the left properly measures the *total* energy loss in the metal, but the integrand is not the *loss density*. It can on the contrary take large positive as well as negative values even when there is no loss at all, i.e. for $\gamma = 0$.

In Sect. 3.10 we discuss that $\text{Re}\{\mathbf{E}^* \cdot \mathbf{j}\}$ integrated only over a region near the surface, where the integrand in the last term of (2.43) has not yet vanished, gives wrong results when interpreted as absorption density. The correct energy loss density is the second integrand in (2.43) as stated in (2.42).

When there are no longitudinal fields, i.e. $\beta = 0$, one needs not to worry. Only in this case the integrands of the first and second integral in (2.43) are equal within the HD.

2.6 The Additional Boundary Conditions at an Interface Between Two Metals of Different Electron Density

When two different metals or two model electron gas systems of different density have a common interface, the problem of transmission and reflection of waves is a bit more complicated than at a free surface. This situation is sketched in Fig. 2.2. Both media are now spatially dispersive and can sustain longitudinal plasma waves as well as transversal waves. Therefore the general solution (2.21) has now two free amplitudes on each side. In total four boundary conditions are necessary to define the solution. The incoming wave can be either transverse or longitudinal (or a combination of both), so here also the problem of reflection and transmission of plasmons at density steps is included.

FORSTMANN and STENSCHKE /2.6 , 28/ have proposed an additional boundary condition for this case. The concepts explained in Sect. 2.4 for the free surface, the non singular nature of charge and current densities, are employed here, too. This again leads to continuity of the normal component of the electric field E_n (2.31b) and of the current j_n, generalizing (2.31c). In addition the continuity of the normal component of the energy current is assumed as an essential requirement. This means, that there is no singular sink or source of energy in the surface plane. If we specify our model of the interface once more as being a plane between homogeneous regions, where all the material parameters n_0, ω_p, β, γ change discontinuously by a finite step and require in addition, that all variables in our equation $\mathbf{E}$, $\mathbf{B}$, ρ, $\mathbf{j}$ are finite, the continuity of the normal energy current can be derived from (2.34)

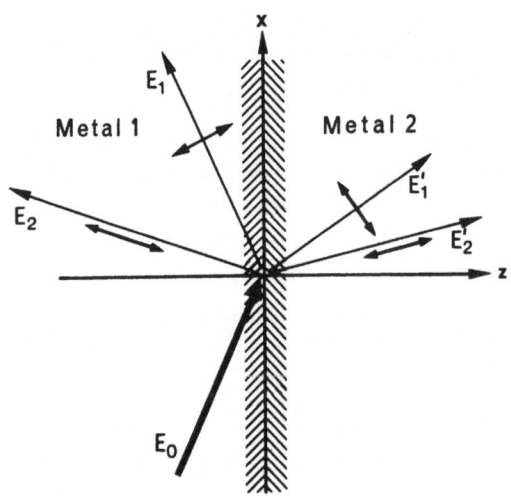

by the usual pillbox argument.

Of the two contributions to the energy current in (2.34), the normal component of the Poynting vector is already continuous, due to the continuity of E_{tg} and B_{tg} (equivalent to continuity of D_n). Therefore the additional boundary condition is:

$$Q_n = \frac{4\pi\beta}{\omega_p^2} \rho \, j_{normal} \quad \text{continuous} . \tag{2.44}$$

Since the normal component of j is already continuous (due to the finiteness of ρ) the second factor must be continuous. We collect the four boundary conditions at an interface between two metals /2.6/:

$$E_{tangential} \quad \text{continuous} \tag{2.45a}$$

$$E_{normal} \quad \text{continuous} \tag{2.45b}$$

$$j_{normal} = \sigma_\perp E_{\perp n} + \sigma_\| E_{\| n} \quad \text{continuous} \tag{2.45c}$$

$$\frac{4\pi\beta}{\omega_p^2} \rho = \frac{\beta}{\omega_p^2} \text{div } E \quad \text{continuous} . \tag{2.45d}$$

Again $D_n = E_n - (4\pi/i\omega)j_n$ is continuous by virtue of (2.45b) and (2.45c). At a free surface (vacuum on one side) the material term Q_n (2.44) of the normal energy current is continuous without (2.45d) because $j_n = 0$, therefore the three conditions (2.31) suffice. In Sect. 3.11 we comment on a different proposal of ABC's by BOARDMAN and RUPPIN /2.24/.

2.7 Extension to Not-Nearly-Free Electron Metals

The metal most often used to investigate the consequences of nonlocal optics is silver, because its plasma frequency defined by $Re\{\epsilon(\omega, k = 0)\} = 0$ is especially low in the near ultraviolet at $\hbar\omega_p = 3.8$ eV ($\lambda = 3263$ Å, $1/\lambda = 30646$ cm^{-1}, $\nu = 9.18 \cdot 10^{14}$ Hz) and easily accessibly by standard optical equipment. In addition the damping of the plasma waves is relatively weak because the interband absorption starts at 4 eV. Therefore the collective modes are well developed.

Silver is not a "simple" free electron metal, it has s and d electrons in its valence shell. The theory developed previously is based on the model of free electrons and is appropriate for simple metals like Al, Na, K, perhaps Ga, In, Tl, Hg etc. How to adapt it to silver? The following approximate concept has been proven successful /2.29 , 30/.

The electrons in silver are considered as belonging to two separated systems, nearly free s-conduction electrons and "bound" d electrons. The density of "free" electrons is determined from the Drude behaviour of $\epsilon(\omega)$ in the infra red /2.31/ which yields one free conduction electron per silver atom. This free electron density or the related plasma frequency ω_n is a parameter for the separation. The two polarizable systems are coupled only via the mean fields treated in Maxwell's equations. This concept is applied in the papers quoted above. The measured dielectric function /2.31/ is separated into free and bound electron part:

$$\epsilon(\omega) = \epsilon_b(\omega) - \frac{\omega_n^2}{\omega(\omega + i\gamma)} = 1 + 4\pi\chi_b + 4\pi\chi_f \quad . \tag{2.46}$$

The free electrons are moving in a polarizable, screening background. The actual plasma resonance $\epsilon(\omega) = 0$ is at $\omega_p^2 = \omega_n^2/\epsilon_b(\omega_p)$ which for silver is at $\hbar\omega_p = 3.8$ eV instead of $\hbar\omega_n = 9.0$ eV prescribed by the density of free conduction electrons.

Only the conduction electron system is now assumed to be spatially dispersive like an electron gas, while spatial dispersion in the background system of bound electrons is neglected. For longitudinal fields the polarizabilities therefore are

$$\chi_{b\parallel}(\omega) = \chi_{b\perp}(\omega) \tag{2.47}$$

and from equation (2.15) for the free s electrons:

$$4\pi\chi_{f\parallel}(\omega, k) = -\frac{\omega_n^2}{\omega(\omega + i\gamma) - \beta k^2} \tag{2.48}$$

$$\epsilon_{\parallel}(\omega, k) = \epsilon_b(\omega) - \frac{\omega_n^2}{\omega(\omega + i\gamma) - \beta k^2} \quad . \tag{2.49}$$

The measured imaginary part in $\epsilon_b(\omega)$ is large and provides enough damping of the plasma waves. Therefore the additional damping parameter γ in (2.48) or (2.49) can often be dropped. One has to express in (2.15) ρ by $\mathbf{j}$ from the continuity equation for the separated conduction electron system or by $\text{div}(\varepsilon_b\mathbf{E}) = \rho$ for the conduction electron charge density in the polarizable background. By either route one arrives at the plasmon dispersion:

$$\omega^2 = \omega_n^2/\varepsilon_b(\omega) + \beta(k_x^2 + n^2) \tag{2.50}$$

which tells us that only the electric force in (2.15) is decreased by the screening of the background but the pressure force of the conduction electron gas, which is responsible for spatial dispersion, is unaffected by the background. This is our model assumption. With such a clear separation of a spatially dispersive electron system in front of a background with a local polarizability it is straight forward to transfer the boundary treatment of the previous sections to this more complicated metal. Maxwell's equations (2.1) and (2.3) are now written as

$$\nabla \times \mathbf{B} = \frac{1}{c}\frac{\partial \mathbf{D}^b}{\partial t} + \frac{4\pi}{c}\mathbf{j} \quad , \quad \nabla \cdot \mathbf{D}^b = 4\pi\rho \tag{2.51}$$

with $\mathbf{D}^b(\omega) = \varepsilon_b(\omega)\mathbf{E}(\omega)$ and ρ and $\mathbf{j}$ only in the conduction electron system. The boundary conditions (2.45) are modified to

$$E_{\text{tangential}} \quad \text{continuous} \tag{2.52a}$$

$$D^b_{\text{normal}} = \varepsilon_b(\omega)E_n \quad \text{continuous} \tag{2.52b}$$

$$j_{\text{normal}} = \frac{i\omega_n^2}{4\pi\omega}E_{\perp n} + \frac{i\omega}{4\pi}\varepsilon_b(\omega)E_{\parallel n} \quad \text{continuous} \tag{2.52c}$$

$$\frac{4\pi\beta}{\omega_n^2}\rho = \frac{\beta\varepsilon_b(\omega)}{\omega_n^2}\text{div }\mathbf{E} \quad \text{continuous} \tag{2.52d}$$

for a system with a background of bound electrons. Equations (2.52) reduce to (2.45) for simple metals with $\varepsilon_b = 1$.

In this section we have described a model which allows the transfer of the concepts derived for free electron metals to more complicated metals. The conduction electron system is spatially dispersive like an electron gas. It carries currents and charge densities which are finite and therefore plasma waves are excited in this system at a surface. The bound electrons on the other hand have a local response and singular surface charges by virtue of (2.52b). This is at least a consistent model, which may be motivated by imagining the charge perturbations in the bound electron system more sharply confined to the surface than in the conduction electron system. But the actual guide was the idea to stay as near to a free electron system as possible and try this pragmatic approach before turning to greater complications. We

20

are aware of the approximations. The application of this model gives qualitatively good results (see Chap. 3), while there is room for quantitative improvement. We are not aware of any other method, which is as able to treat optical problems for silver near its plasma frequency.

3. Applications of Nonlocal Metal Optics

In Chap. 2 an extension of metal optics was discussed, which takes into account that longitudinal plasma waves can exist in a metal because the longitudinal dielectric function depends on the wavevector $\mathbf{k}$, which means a nonlocal relation between field and current. In principle this extension is necessary for all frequencies but in practice it is important for frequencies around the plasma frequency. In this chapter we discuss examples of its application. These examples cover a frequency range from below $\omega_p/\sqrt{2}$ to frequencies above ω_p, where the plasmon wavelength comes close to interatomic distances. At very low frequencies, for several optical questions the singular surface charge of standard optics might be a good approximation. At higher frequencies far above ω_p the plasma waves are no longer well defined modes and the hydrodynamic approximation (2.15) looses its validity. One cannot treat effects of spatial dispersion at very high frequencies on a macroscopic level. Our method and examples cover the intermediate range.

3.1 Reflection and Transmission

Obviously the first effect following from the inclusion of plasma waves in optics is a change in reflectivity from a metal surface or in transmission through metal films. These effects were already studied by SAUTER and coworkers /2.2 , 3.1 , 2/. In order to demonstrate the application of the nonlocal optics we calculate the reflection and transmission amplitude at a vacuum/metal surface for p-polarized light.

According to Fig. 2.1 and formulas (2.21 - 30) we have to match the general solution outside ($\omega_p = 0$)

$$\mathbf{E}(\mathbf{r},t) = e^{i(k_x x - \omega t)} (\mathbf{E}_0 e^{i\lambda_0 z} + \mathbf{E}_r e^{-i\lambda_0 z}) \tag{3.1}$$

to the general solution inside

$$E(\mathbf{r},t) = e^{i(k_x x - \omega t)} (E_1 e^{i\lambda z} + E_2 e^{i\eta z}) \tag{3.2}$$

by the boundary conditions (2.31):

$$E_{tg} \text{ continuous:} \quad \lambda_0 E_r + \lambda E_1 - \frac{k_x^2}{\eta} E_2 = \lambda_0 E_0 \tag{3.3}$$

$$E_n \text{ continuous:} \quad -E_r + E_1 + E_2 = E_0 \tag{3.4}$$

$$j_n \text{ continuous} = 0: \quad \sigma_\perp E_1 + \sigma_\parallel E_2 = 0 \quad . \tag{3.5}$$

The unknowns in our formulation are the z components of the fields (2.28 , 29), not the moduli of the vectors. The reflection amplitude R_p is

$$R_p = \frac{E_r}{E_0} = \frac{\varepsilon\lambda_0 - \lambda + (\varepsilon - 1)k_x^2/\eta}{\varepsilon\lambda_0 + \lambda - (\varepsilon - 1)k_x^2/\eta} \quad . \tag{3.6}$$

The amplitude of the transmitted transversal wave is

$$E_1 = \frac{2\lambda_0}{\varepsilon\lambda_0 + \lambda - (\varepsilon - 1)k_x^2/\eta} E_0 \tag{3.7}$$

and the amplitude of the transmitted longitudinal wave is

$$E_2 = \frac{2\lambda_0(\varepsilon - 1)}{\varepsilon\lambda_0 + \lambda - (\varepsilon - 1)k_x^2/\eta} E_0 \quad . \tag{3.8}$$

The spatial dispersion is switched off by $\beta = 0$ in (2.15), i.e. $\eta \to \infty$ in (2.27). In this limit (3.6) and (3.7) lead to Fresnel's reflection and transmission coefficients:

$$r_{Fr} = \left|\frac{E_r}{E_0}\right|^2 = \left|\frac{\varepsilon\lambda_0 - \lambda}{\varepsilon\lambda_0 + \lambda}\right|^2 \tag{3.9}$$

$$t_{Fr} = Re\left(\frac{\lambda\varepsilon^*}{\lambda_0}\right)\left|\frac{E_1}{E_0}\right|^2 = Re\left(\varepsilon^*\frac{\lambda}{\lambda_0}\right)\left|\frac{2\lambda_0}{\varepsilon\lambda_0 + \lambda}\right|^2 \quad . \tag{3.10}$$

Near ω_p and above, the additional terms in (3.6) reduce the reflection coefficient, as seen in Fig. 3.1. The differences in reflection due to plasma waves are small. They are only visible in the $\log(1 - r)$ plot and do not exceed 1,5 % of r_{Fr}.

We see from (3.7 , 8) that near ω_p ($\varepsilon \approx 0$) the amplitudes of the transverse and longitudinal wave are of the same order of magnitude. The small influence on the reflectivity is caused by the small energy current carried by the plasmon, which is smaller by a factor $v_F/c \approx 10^{-2}$ than the energy current in the transverse wave.

At the surface the normal component of the electric field E_z is (outside and inside)

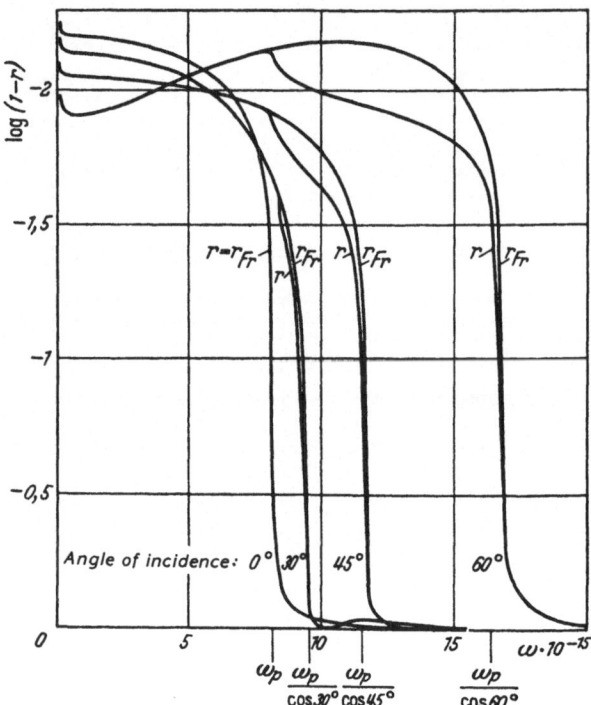

Fig. 3.1. The reflection coefficient r, calculated by nonlocal optics for different angles of incidence; r_{Fr} from local optics (for Na: $\omega_p = 8.2 \cdot 10^{15}$ s^{-1}, $\gamma/\omega_p = 3 \cdot 10^{-3}$, $v_F = 9.87 \cdot 10^7$ cm/s) /3.1/

$$E_0 + E_r = E_1 + E_2 = \frac{2\lambda_0 \epsilon}{\epsilon\lambda_0 + \lambda - (\epsilon - 1)k_x^2/\eta} E_0 \qquad (3.11)$$

while in standard optics we get

$$\text{outside:} \quad E_0 + E_r = \frac{2\lambda_0 \epsilon}{\epsilon\lambda_0 + \lambda} E_0 \qquad (3.12)$$

$$\text{inside:} \quad E_1 = \frac{2\lambda_0}{\epsilon\lambda_0 + \lambda} E_0 \ . \qquad (3.13)$$

The field outside the metal is very little changed by spatial dispersion and goes to zero at ω_p, where $\epsilon = 0$. But the inside field is different. It continuously joins to the outside field for nonlocal optics and so also goes to zero at ω_p, but in standard optics, there is a jump in the electric field and E_z (inside) stays large also at ω_p. This has important consequences for photoemission (see Sects. 3.9, 10). The decay length (Im η)$^{-1}$ is much shorter than the decay length (Im λ)$^{-1}$ for the transverse wave. Deep inside the metal only E_1 (3.7) survives and differs then usually very little from the value calculated by standard optics. Important effects appear, when in a layer another boundary is met before the plasma wave contribution to the fields has vanished. An example of this enhancement of nonlocal effects in thin layers can be seen in Fig. 3.2. The reflection and transmission coefficients

24

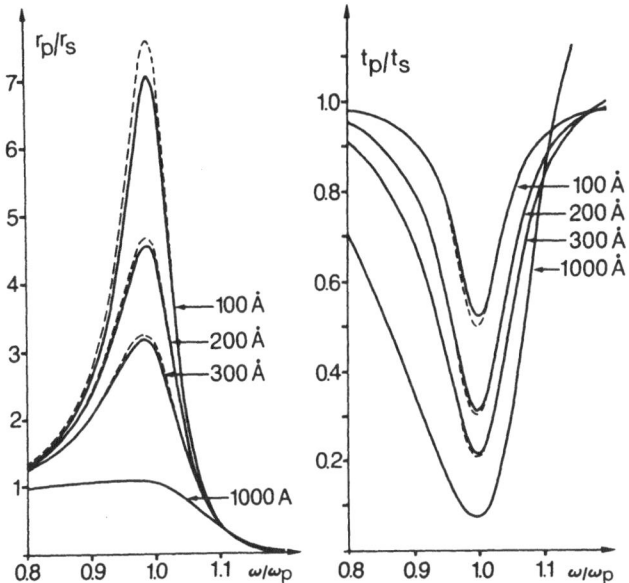

Fig. 3.2. The quotient r_p/r_s of the reflection coefficients and t_p/t_s of the transmission coefficients for p- and s-polarized light for different metal layer thickness. Calculation including plasma waves (———) and without plasma waves (-----) /3.2/

for p-polarized light devided by those for s-polarized light are shown for metal layers of different thickness /3.2/. The resonance character near ω_p is already present in Fresnel's formulas for p polarization and is not due to "plasma wave excitations"; but coupling of plasma waves to the light opens another channel for energy transmission and therefore increases the energy transmission through the layer and decreases the reflection for p-polarized light, if the layers are thin enough for the plasma wave to reach the second boundary.

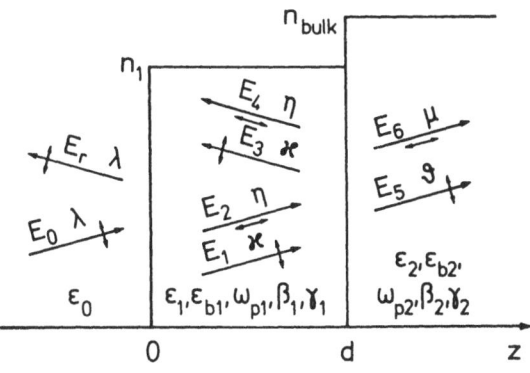

Fig. 3.3. The fields and parameters for a two step surface model

In later sections we study the influence of a layer of lower electron density on top of the semiinfinite bulk density metal, Fig. 3.3, in order to investigate the influence of a soft decrease of the electron density at a metal surface or to calculate the response of deposited metal layers. For the system of Fig. 3.3 we have three boundary conditions (2.31) at $z = 0$ and 4 conditions (2.45) at $z = d$ which yield 7 linear equations for the 7 unknown amplitudes. We give the reflection amplitude for the system of Fig. 3.3 for an ansatz analogous to (3.1-6) and include a background dielectric function ε_0, ε_{b1}, ε_{b2} (Sect. 2.7):

$$R_p = E_r/E_0 = NUM/DEN \tag{3.14}$$

$$NUM = DP \cdot CMM + DM \cdot CPM \cdot a^2 + FP \cdot CMP \cdot b^2 + FM \cdot CPP \cdot a^2 \cdot b^2 - H \cdot a \cdot b \tag{3.15}$$

$$DEN = DP \cdot CPP + DM \cdot CMP \cdot a^2 + FP \cdot CPM \cdot b^2 + FM \cdot CMM \cdot a^2 \cdot b^2 + H \cdot a \cdot b \tag{3.16}$$

$$a = \exp(-\kappa d) \quad , \qquad b = \exp(-\eta d) \tag{3.17}$$

$$D_M^P = (1 + \frac{\eta}{\mu} \delta) B_M^P + G \cdot B_{MM}^{PP} \tag{3.18}$$

$$F_M^P = (1 - \frac{\eta}{\mu} \delta) B_M^P + G \cdot B_{MM}^{PP} \tag{3.19}$$

$$H = 8\kappa k^2 (\varepsilon_2 \delta - \varepsilon_1)/\mu \quad , \qquad G = \frac{\varepsilon_{b2}}{\varepsilon_2} \frac{\varepsilon_2 \delta - \varepsilon_1}{\varepsilon_1 - \varepsilon_{b1}} \tag{3.20}$$

$$C_{MM}^{PP} = \lambda \varepsilon_1/\varepsilon_0 \pm \kappa \pm k^2 (\varepsilon_1 - \varepsilon_{b1})/(\eta \varepsilon_{b1}) \tag{3.21}$$

$$B_{MM}^{PP} = \kappa \varepsilon_2 \pm \vartheta \varepsilon_1 \pm k^2 (\varepsilon_2 \delta - \varepsilon_1)/\mu \tag{3.22}$$

$$B_M^P = \kappa \varepsilon_2 \pm \vartheta \varepsilon_1 \tag{3.23}$$

$$\varepsilon_j = \varepsilon_{bj} - \omega_{pj}^2/\omega(\omega + i\gamma_j) \quad , \qquad \delta = \varepsilon_{b1}/\varepsilon_{b2} \tag{3.24}$$

$$\lambda = (k_x^2 + k_y^2 - \varepsilon_0 \omega^2/c^2)^{1/2} \tag{3.25}$$

$$\kappa = (k_x^2 + k_y^2 - \varepsilon_1 \omega^2/c^2)^{1/2} \tag{3.26}$$

$$\vartheta = (k_x^2 + k_y^2 - \varepsilon_2 \omega^2/c^2)^{1/2} \tag{3.27}$$

$$\eta = [k_x^2 + k_y^2 - \varepsilon_1 \omega^2 (1 + i\gamma_1/\omega) / (\varepsilon_{b1} \beta_1)]^{1/2} \tag{3.28}$$

$$\mu = [k_x^2 + k_y^2 - \varepsilon_2 \omega^2 (1 + i\gamma_2/\omega) / (\varepsilon_{b2} \beta_2)]^{1/2} \tag{3.29}$$

$$\text{Re}\{\lambda,\kappa,\vartheta,\eta,\mu\} > 0 \quad , \qquad \text{Im}\{\lambda,\kappa,\vartheta,\eta,\mu\} < 0 \quad . \tag{3.30}$$

The denominator of this reflection amplitude will be used in a discussion of the higher surface modes (Sect. 3.4). The reflectivity of a variety of systems can be calculated by taking limits of (3.14). Spatial dispersion can be neglected by $\beta \to 0$, $\eta \to \infty$. But if one wants a metal layer on a non spatially dispersive substrate the limit has to be taken a bit more carefully: $\varepsilon_2 \to \varepsilon_{b1}$; $\varepsilon_{b2} \to \varepsilon_{b1}$ everywhere except in (3.27) and in the transition $\vartheta \to (\varepsilon_{b1}/\varepsilon_2)\vartheta$. Now $\lim\limits_{\mu \to \infty} R_p$ yields the reflection amplitude from the terms linear in μ^{-1}.

We add here, that for s-polarized light, where the electric field has only a component parallel to the surface, the problem of spatial dispersion, i.e. of plasma wave excitation disappears and standard optics applies. For a one step surface (Fig. 2.1) we get:

$$R_s = E_{ry}/E_{0y} = (\lambda_0 - \lambda)/(\lambda_0 + \lambda) \tag{3.31}$$

$$T_s = E_{1y}/E_{0y} = 2\lambda_0/(\lambda_0 + \lambda) \quad . \tag{3.32}$$

Physically this can be understood, because tangential fields will not induce any charge density perturbations near the surface. Therefore div $\mathbf{E}$ = 0 in the metal and only transverse fields play a role. Mathematically, when the plane of incidence is the x , z plane, longitudinal waves parallel to $\mathbf{k}_L = (k_x,0,\eta)$ do not have a y component. Because the boundary conditions for the fields in the y direction are decoupled from those for the x , z fields (see also Chap. 4), the x , z components become zero for s-polarized light. Consequently also the plasma wave amplitude is zero. From the material equation no new boundary condition appears, because j_n is identically zero. Therefore, if there is an uncertainty about the influence of spatial dispersion in one case or another, an investigation with s-polarized light or normal incidence can reduce the complication (e.g. Sect. 3.6). In certain semiconductors also for s polarization spatial dispersion plays a role /2.4 , 9 , 10; 3.3 , 4/, because the transverse dielectric function can be k dependent, too (see Sect. 4.2). In a hydrodynamic model this is equivalent to having shear forces in the system /2.10/. In view of the experiments it is a good approximation for the conduction electrons in the metal, to treat them according to (2.15) as an ideal liquid without shear forces.

3.2 Resonances in Thin Metal Films

MELNYK and HARRISON /3.5/ realized, that the effect of the additional plasma waves might become detectable, when they form standing waves in thin layers, i.e. when resonant conditions are achieved. They predicted minima in transmission and maxima in reflection and absorption whenever odd multiples of the normal component of half the plasma wavelength λ_L fit into the thickness d of the film:

$$n(\lambda_L)_{normal} = 2d \quad , \quad n = 1,3,5 \tag{3.33}$$

or with (2.27)

$$\eta d = \pi n \quad n = 1,3,5 \quad . \tag{3.34}$$

These resonances where beautifully demonstrated by LINDAU and NILSSON /3.6/, who detected the minima in transmission through a silver foil of 120 Å thickness, Fig. 3.4,

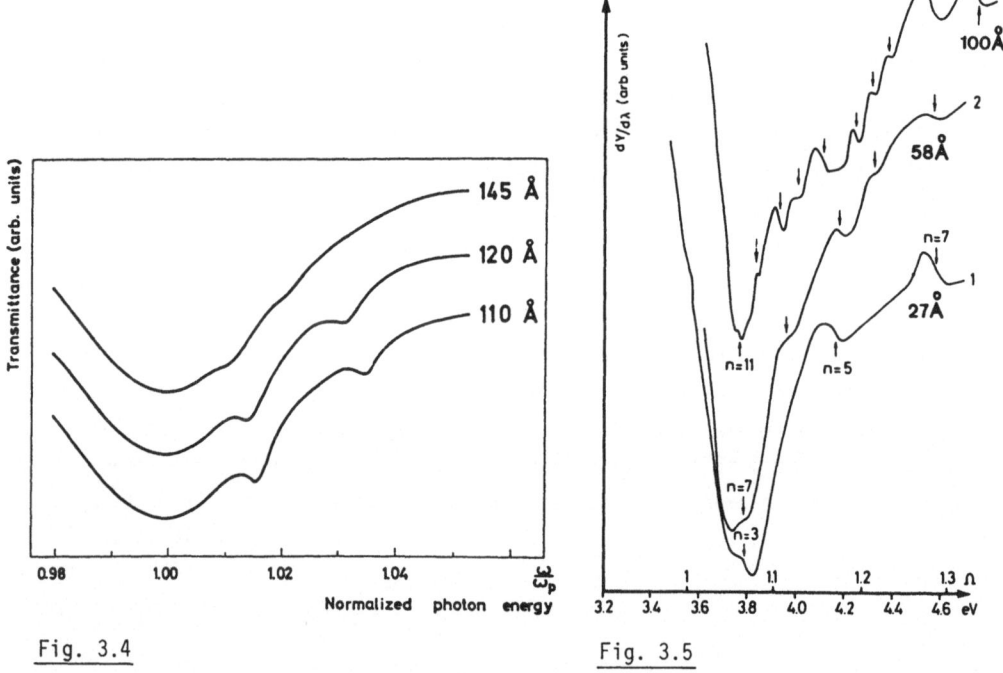

Fig. 3.4   Fig. 3.5

Fig. 3.4. Transmittance spectra of Ag films of three different thicknesses for p-polarized light at 75° incidence angle /3.6/

Fig. 3.5. Wave length modulated photoemission yield spectra $\partial Y/\partial\lambda$ from potassium films of three different thicknesses. p-polarized light at 45° incidence angle /3.7/

and by ANDEREGG, FEUERBACHER, FITTON /3.7/, who measured the resonance maxima
in absorption by means of maxima in the total photoemission yield from potassium
films of 100 Å and less thickness, Fig. 3.5. In both papers the plasmon dispersion
was evaluated from the sequence of resonances and very close agreement to the qua-
dratic behaviour (2.20) was found. This evaluation was later debated /3.8/ but there
were no recent attempts to check these experiments. The relation between the absorption and the photoemission yield is discussed in more detail in Sect. 3.10. Both
quantities are assumed proportional to each other.

One may wonder, why only the odd multiples of $\lambda/2$ lead to a resonance. This fact
depends on the boundary conditions (2.31). The analysis of fields and currents at
the frequencies, where odd or even multiples of half the plasma wave length fit into the layer thickness, shows that for odd multiples the forward and backward running
plasma waves interfere constructively, while for even multiples, they wipe out each
other. This choice of phase for the reflected wave can be related to $j_n = 0$ at the
surface. The current is the sum of a transverse and a longitudinal contribution
(2.24). The transverse current has the same sign at both surfaces due to the very
large transverse wave length. If an even number of $\lambda_p/2$ fits into the layer, the
longitudinal current changes sign from one surface to the other. $j_n = 0$ is only
achievable, if the plasma waves cancel each other as well as the transverse waves
do (destructive interference). For odd multiples of $\lambda/2$, the current in the plasma
waves has the same sign at both surfaces and can compensate a nonvanishing transverse current. In this case the boundary condition $j_n = 0$ can be fulfilled even
when the reflected plasma wave interferes constructively with the incoming longitudinal wave as it does. Large longitudinal current densities occur leading to large
absorption according to (2.42), to larger reflection and small transmission. A dependence of these results on different choices of boundary conditions is discussed
in Sect. 3.11.

3.3 The Surface Plasmon Dispersion

The discussion of the surface plasmon can be presented as a problem of optics:
What are the eigenmodes of the system of fields, charges and currents near a metal
surface? Does the reflection problem (3.3 - 5) have nonzero outgoing solutions for
vanishing amplitude E_0 of the incoming wave?

Since the discovery of the surface plasmon eigenmode of the bounded electron
gas by RITCHIE /3.9/ a large number of articles studied this mode in the framework
of many body theory. The complications of this electron gas treatment usually do
not allow to treat also the electromagnetic fields carefully. Therefore Maxwell's
equations are often curtailed to div $\mathbf{E} = 4\pi\rho$, i.e. effectively to electrostatics
with the velocity of light $c \to \infty$.

The other extreme approximation reduces the electron response to a local dielectric function and searches for eigenmodes in the framework of standard optics. The eigenmode found in this way is sometimes called a surface polariton or a surface plasmon polariton. We will outline here that the nonlocal optics discussed in Chap. 2 can easily bridge this gap between the two approaches and that one can learn the essential physics of the surface plasmon by use of the HD approximation. There exists a number of reviews about surface plasmons /2.9 , 3.10 - 13/, especially /3.12/ contains an extended list of references. Here only those are listed, which are explicitly quoted.

We first look for the surface eigenmode within standard optics. For the reflection of p-polarized light at a vacuum-metal interface the boundary conditions (3.3 - 5) in the case of local optics reduce to:

$$
\begin{array}{l} E_{tg} \text{ contin.:} \\[1em] D_n \text{ contin.:} \end{array}
\begin{pmatrix} \lambda_0 & \lambda \\ -1 & \varepsilon \end{pmatrix}
\begin{pmatrix} E_r \\ E_1 \end{pmatrix}
=
\begin{pmatrix} \lambda_0 \\ 1 \end{pmatrix} E_0 \quad .
\tag{3.35}
$$

The eigenmode with nonzero fields for vanishing excitation requires the determinant of the coefficient matrix to vanish:

$$
\varepsilon \lambda_0 + \lambda = \varepsilon \sqrt{\frac{\omega^2}{c^2} - k^2} + \sqrt{\frac{\omega^2}{c^2}\varepsilon - k^2} = 0 \quad .
\tag{3.36}
$$

This dispersion relation has been derived by ZENNECK /3.14/ and SOMMERFELD /3.15/ in connection with radio wave propagation along the surface of the sea. Its relation to surface plasmons was pointed out by STERN /3.16/ and OTTO /3.17/. In Fig. 3.6 this dispersion is labelled 'Zenneck'. For small k the frequency goes to zero as:

$$
\omega_s = c k \left(1 - \frac{1}{2}\frac{c^2 k^2}{\omega_p^2}\right) \quad ; \qquad c k = \omega_s \left(1 + \frac{1}{2}\frac{\omega_s^2}{\omega_p^2}\right)
\tag{3.37}
$$

and for $k \to \infty$

$$
\varepsilon(\omega_s) \to -1 \quad , \qquad \omega_s \to \omega_p/\sqrt{2}
\tag{3.38}
$$

when $\varepsilon = 1 - \omega_p^2/\omega^2$.

The z components λ_0 and λ (2.25 , 26) of the wave vectors in the general solutions of Maxwell's equations are positive imaginary at ω_s (if the damping γ is neglected); the fields decay exponentially at both sides of the surface. The eigenmodes cannot couple to incident waves because for a given frequency its wavelength along the surface is shorter, k is larger than possible for a light wave, as can also be seen in (3.37). The technique of attenuated total reflection (ATR) /3.18 , 19/ overcomes

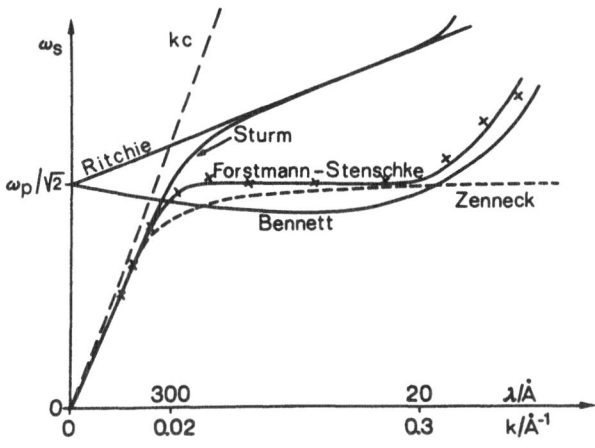

Fig. 3.6. The surface plasmon dispersion in different approximations (schematic). (x x x) indicate the experimentally found dispersion for Al

this problem by reducing the wavelength of light in a glasprism above the metal surface so that frequency and wavevector can fulfill (3.36) and the surface plasmons can be studied optically. Relation (3.36) is often squared and solved for k^2 as

$$k^2 = \frac{\varepsilon}{\varepsilon + 1} \frac{\omega_s^2}{c^2} \quad . \tag{3.39}$$

This equation has in addition to the solution, which in Fig. 3.6 is denoted 'Zenneck', a solution starting at $k = 0$, $\varepsilon = 0$, which means $k = 0$, $\omega = \omega_p$, and approaching $k^2 = (1/2)\omega^2/c^2$ for large frequencies. This solution is sometimes called the 'radiative surface mode'. It was recently pointed out again by MARADUDIN /3.12/ that this solution of (3.39) does not solve (3.36), because the ansatz (3.1,2) has the consequence that the roots have both positive real parts and positive imaginary parts, so for $\omega > \omega_p$ with $\varepsilon > 0$ (3.36) cannot have a solution. The upper solution is not a condition for nonvanishing fields without external excitation, but instead predicts under which condition there is *no reflected wave* when a wave comes in. This is the condition for the Brewster angle. From (3.35) it can be seen, that a change of the sign in (3.36) gives the Brewster condition $E_r = 0$, not an eigenmode.

Up to now our discussion of the surface plasmon dispersion did not really involve the plasma. But the surface plasmon is imagined as a charge density wave travelling along the surface. Standard optics keeps the interior of the metal free of charge and allows charge only in a singular surface plane. Therefore the treatment of the surface plasmon is the most obvious field where the nonlocal optics should be applied. This was actually done by STURM /3.20/ immediately after the plasma wave optics was proposed.

Before discussing this extension, we show, that the equation of motion (2.15) contains the physics necessary to derive the surface plasmon dispersion which is obtained by most many body treatments. In the electrostatic approximation the equation of motion (2.15) is supplemented by Possion's equation for the potential φ and the continuity equation. Eliminating $\mathbf{E}$ and $\mathbf{j}$ from (2.15) and with the ansatz $\exp(ikx - i\omega t)$ for the x and t dependence of all quantities, we obtain (dropping the damping in (2.15)):

$$\beta\left(\frac{\partial^2}{\partial z^2} - k^2\right)\rho = (\omega_p^2 - \omega^2)\rho \qquad z > 0 \tag{3.40}$$

$$\left(\frac{\partial^2}{\partial z^2} - k^2\right)\varphi = \begin{cases} -4\pi\rho & z > 0 \\ 0 & z < 0 \end{cases}. \tag{3.41}$$

The general solutions

$$\rho(z) = \rho_0 e^{-\kappa z} \quad , \qquad \kappa = \left[\frac{1}{\beta}(\omega_p^2 - \omega^2) + k^2\right]^{1/2} \tag{3.42}$$

$$\varphi(z) = \frac{4\pi}{k^2 - \kappa^2}\rho_0 e^{-\kappa z} + A e^{-kz} \qquad z > 0 \tag{3.43}$$

$$\varphi(z) = B e^{kz} \qquad z < 0 \tag{3.44}$$

have to be matched at the surface $z = 0$ by the three boundary conditions

continuity of φ: $B = A + \dfrac{4\pi}{k^2 - \kappa^2}\rho_0$,   (3.45)
(from continuity of E_x)

continuity of $E_z = -\dfrac{\partial\varphi}{\partial z}$: $-kB = kA + \dfrac{4\pi\kappa}{k^2 - \kappa^2}\rho_0$,   (3.46)

continuity of the normal component of the current density j_z, which from the z component of (2.15) yields

$$-\frac{\omega_p^2}{4\pi}\frac{\partial\varphi}{\partial z} - \beta\frac{\partial\rho}{\partial z} = 0 \qquad \text{at } z = 0 \text{ , i.e.} \tag{3.47}$$

$$\frac{\omega_p^2}{4\pi}\left(kA + \frac{4\pi\kappa}{k^2 - \kappa^2}\rho_0\right) = -\kappa\beta\rho_0 \quad . \tag{3.48}$$

A solution exists, if (3.45 , 46 , 48) are linear dependent, which yields

$$-\frac{\omega_p^2}{2}\frac{1}{k + \kappa} = -\kappa\beta \tag{3.49}$$

$$\omega_s^2 \approx \frac{\omega_p^2}{2}\left(1 + \frac{k}{\kappa}\right) = \frac{\omega_p^2}{2} + k\sqrt{\beta}\frac{\omega_p}{\sqrt{2}} \tag{3.50}$$

$$\omega_s \approx \frac{\omega_p}{\sqrt{2}} + \frac{k}{2}\sqrt{\beta} \tag{3.51}$$

Equation (3.51) or (3.50) give the dispersion for the surface plasmon in the electrostatic, 'unretarded' approximation ($c \to \infty$) as derived by RITCHIE /3.9/ and most other papers concerned with this eigenmode (see /3.12/). In the limit $k \to 0$, the frequency is finite, in this approximation, $\omega_s = \omega_p/\sqrt{2}$, and the dispersion starts linearly in k, see 'Ritchie' in Fig. 3.6. The coefficient of the term linear in k has been discussed in several papers (see /3.12/). As we show later, it depends on the density gradient of the electron gas near the surface. In the electrostatic approximation the fields outside of the metal are curl-free but oscillate in time, whereby they indicate the inconsistency of the electrostatic treatment.

An even more radical approximation combines Poisson's equation with a local dielectric function, keeps the metal free of charge and allows only for singular surface charges. Then the homogeneous parts of the solutions (3.43 , 44) are matched by φ continuous, $D_n = -\varepsilon(\partial\varphi/\partial z)$ continuous, which is only possible for

$$\varepsilon_1 = -\varepsilon_2 \quad . \tag{3.52}$$

For the interface between two metals this interface mode lies between the lower and the upper plasma frequency of the two metals. HARRIS and GRIFFIN /3.21/ showed in a more sophisticated calculation, that the spreading of the charge perturbation, i.e. a treatment beyond the local optics, is necessary to get the linear dispersion (3.51).

It is obvious that the eigenmode derived by nonlocal optics will combine the consequences of a treatment of the electron gas in the hydrodynamic approximation (2.15) with those of a proper treatment of the field equations. We get the dispersion from the determinant of the coefficient matrix of (3.3 - 5) or the denominator in (3.6) being zero /3.20/:

$$\varepsilon\lambda_0 + \lambda - (\varepsilon - 1)k_x^2/\eta = 0 \quad . \tag{3.53}$$

The symbols are given by (2.25 - 27 , 30). The graph of (3.53) is denoted 'Sturm' in Fig. 3.6. η is of the order ω_p/v_F for frequencies well below ω_p, while $k \sim \omega/c$. Therefore in the small k, small ω regime (3.53) reduces to (3.36) i.e. the surface plasmon dispersion for long wavelengths is determined mainly by the finiteness of the speed of light, not by electron gas properties. The latter come in at shorter wavelengths, where the electron gas pressure raises ω_s above the value $\omega_p/\sqrt{2}$. The line $\omega = \omega_p/\sqrt{2}$ is crossed for

$$k \approx \frac{\omega}{c}\left(\frac{c}{2\sqrt{\beta}}\right)^{1/3} \tag{3.54}$$

which can be used as a rough boundary between the two regimes. At large k the dispersion follows for some range (3.51) and later ω_s increases stronger than linearly, see Fig. 3.6.

The linear increase is relatively steep, see Fig. 3.7, where the broken curve shows (3.53) for parameters of aluminum. It was soon recognized that the experiments

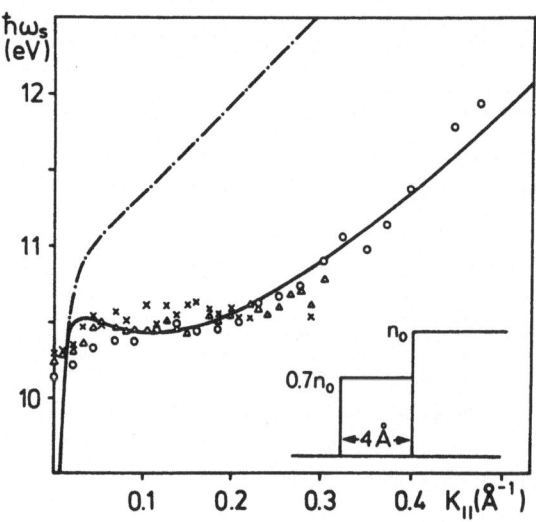

Fig. 3.7. The surface plasmon dispersion. Data points from energy loss measurements /3.22/. (————) is a calculation by non-local optics for a surface model as sketched in the insert, n_0 for Al. (—·—·—) is a graph of (3.53) for Al parameters /2.28/

show a flat plateau around $\omega_p/\sqrt{2}$ ($\varepsilon(\omega) = -1$). In 1976 it was measured /3.22/ that the dispersion increases at all above $\omega_p/\sqrt{2}$ for very short wavelengths. It was shown by BENNETT /3.23/ in the hydrodynamic approximation, and later by several others (for references see /3.12, 13/), that a continuous decrease of the electron density at the surface, a 'soft' boundary, leads to a decrease in the linear term in (3.51), which can even become negative if the transition region, where the electron density drops from the inside value to zero outside, is rather extended. The reason for this decrease in frequency is a 'skin-effect'. From (2.25, 27) it can be seen, that the decay length of the fields inside the metal shrinks when k_x grows (and ω is nearly constant). A shrinking penetration depth in a charge density gradient weighs more and more the outer lower densities. The perturbation sees a lower effective plasma frequency, which also leads to a decrease of the surface plasmon frequency. This decrease with growing k is competing with the increase (3.51) due to the electron gas pressure, and the parameters of the density gradient at the surface determine, which effect wins. We have drawn schematically in Fig. 3.6 a typical dispersion curve for a soft surface in the electrostatic approximation and denoted it 'Bennett'.

Measurements of the surface plasmon dispersion at an Al surface in an extended k range /3.22/ showed the plateau at $\omega_p/\sqrt{2}$ and an increase for large k, see Fig. 3.7. These data were analysed by a calculation using nonlocal optics /2.28/. The soft surface density gradient was approximated by a step of smaller density (Fig. 3.3) because in the approach of phenomenological optics homogeneous regions with planar interfaces are necessary. It is pointed out in Sect. 3.5, 6 that such a system (see sketch Fig. 3.7) shows a separate resonance for each interface when treated by standard optics, because perfect screening by singular surface charges decouples

34

the interfaces. With nonlocal optics the charge density perturbations are spread
and a single eigenmode of the compound system results. The parameters of the surface step have been determined by an optimal fit to the experimental data /2.28/.
Fig. 3.7 shows clearly that the plasma wave optics with a stepped surface model can
describe easily all the important physics in the dispersion curve, the $\lim_{k \to 0} \omega_S = 0$,
the plateau near $\omega_S = \omega_p/\sqrt{2}$ for intermediate k values and the increase of ω_S for
large k. Schematically, the dispersion from nonlocal optics at a 'soft' surface is
shown in Fig. 3.6 under notation 'Forstmann-Stenschke'.

The surface plasmon dispersion measurement was the first experiment known to be
sensitive to the parameters of the surface step and therefore this model of a 'soft
surface' has also been used in other optical studies (Sects. 3.4 - 10). Even the
higher surface modes related to standing waves in the selvedge have recently been
seen experimentally (Sect. 3.4).

The experiment leading to Fig. 3.7 was eléctron energy loss measurement in transmission /3.22/. The small k region is better investigated with optical methods, for
instance ATR /3.18 , 19/. An especially beautiful example is shown in Fig. 3.8, where
even differences for different orientations of a single crystal can be resolved
/3.24/. Usually approximation (3.36) describes the dispersion in this regime sufficiently well. For silver $\varepsilon(\omega) = -1$ for $\hbar\omega = 3.65$ eV [29 436 cm^{-1}].

We discussed here only surface plasmons at a metal half space. The methods de-

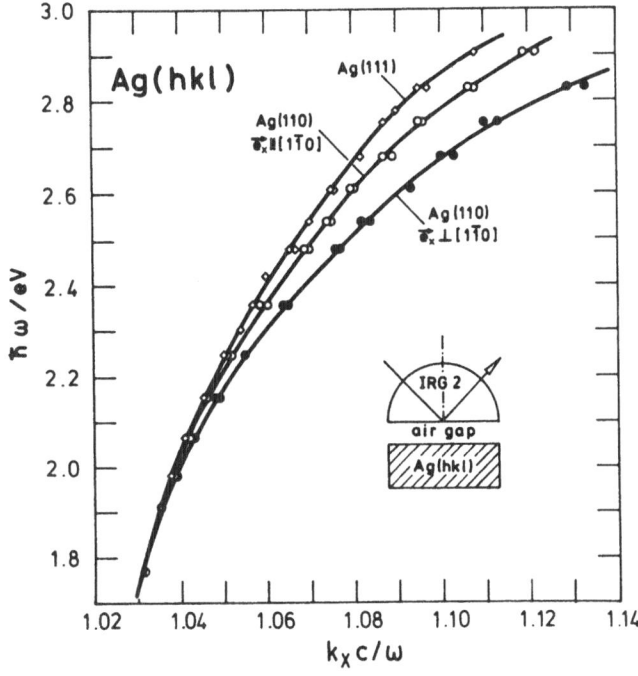

Fig. 3.8. The surface
plasmon dispersion for
small k for different
single crystal surfaces
of silver from ATR mea-
surements /3.24/

scribed here have also been applied to metal films and layered structures. The dispersion of the eigenmodes of a metal layer derived by nonlocal optics was first published in /3.2/. The numerous literature about the eigenmodes of films has been covered in the quoted review articles (e.g. /3.12/).

3.4 Standing Wave Eigenmodes in Thin Surface Layers

When a metal of plasma frequency ω_p is covered with a spatially dispersive surface layer with lower plasma frequency ω_L this surface layer may act like a resonance cavity for plasma waves bound to the surface. The frequency of these modes must be below the bulk ω_p for proper surface modes in order that the tail of these excitations decays exponentially into the bulk. For higher frequencies one can at best have surface resonances. In the frequency range between ω_L and ω_p the number of eigenmodes is determined by the plasmon wavelength in that region, by the geometry of the cavity, i.e. the layer thickness and shape, and by the phase relations at the surface and inner interface, because the eigenmodes are essentially standing plasma waves in the surface layer. The phaseshifts for the inner reflections are determined by the additional boundary conditions and therefore different ABCs yield slightly different eigenmode spectra. If for a given difference $\omega_p - \omega_L$ the thickness of the surface layer grows, more and more resonances drop below ω_p and increase the number of real eigenmodes. The same is true for decreasing ω_L at fixed thickness. The story is quite analogous to bound states in a one dimensional finite potential well and EGUILUZ et al. /3.25/ have mapped the problem in the electrostatic limit onto a one dimensional Schrödinger equation. They called these modes "multipole modes", because in the electrostatic $k \to 0$ limit, the modes don't have a net charge. Boardman pointed out, that the zero charge conditon does not survive, if retardation is included. Because this limit is rather artificial and mathematical anyway, we prefer the term standing wave eigenmodes because of the clear physical picture it expresses.

Every eigenmode has a dispersion depending on k, the wavevector parallel to the surface. The dispersion of these surface modes for a large variety of models has been evaluated /3.23 , 25 - 31/. Also on the surface of a single metal one can expect in principle this kind of higher eigenmodes, because the electron density decays gradually to zero at the surface and therefore every metal surface has effectively a layer of lower electron density on top of the bulk density. BENNETT /3.23/ was the first to recognize, that this low density surface selvedge, which he approximated by a linear decay from the bulk density to zero, can sustain higher resonances in addition to the standard surface plasmon (Sect. 3.3), if the profile is soft enough and if the low density region is broad enough. A series of later papers by

BOARDMAN and collaborators /3.29 - 31/ and by EGUILUZ, QUINN et al. /3.25 - 28/ con-
firmed this result for several density profiles and including retardation. BOARD-
MAN /3.13/ recently reviewed these calculations. The question, which these model
calculations left open, was: Is the density decay on a real metal soft enough, that
a real metal actually sustains a second surface eigenmode or more? Which experiment
can show such a higher mode?

Recently SCHWARTZ and SCHAICH /3.32/ gave a positive answer to these questions.
They pointed out, that the large field near the surface leading to the high photoemission yield from aluminum below ω_p (see Sect. 3.10) is related to the existence
of a first standing wave eigenmode (in addition to the standard surface plasmon) in
the decaying surface density. Their argument is based on the work of KEMPA and
FORSTMANN (/3.33/, see Sect. 3.9) who derived the enhanced field and the large photoyield for a surface model which was fitted to the surface plasmon dispersion of Al
and which shows a higher eigenmode /2.28/. SCHWARTZ and SCHAICH argue that in order
to get the field enhancement the incoming light must couple to the remnants to the
left of the light line of this eigenmode proper. They demonstrate that the large
photoyield disappears for a surface model which is "less soft" and pushes the eigenmode above ω_p into the resonance region. KEMPA and GERHARDTS /3.34/ have worked out
recently, that also a selfconsistent jellium surface sustains such a standing wave
eigenmode for bulk densities comparable to Al (see Sect. 5.6).

The higher surface modes have been experimentally detected recently in a different context, not really as eigenmodes, but as standing waves excited by an incoming light wave. PIAZZA et al. /3.35/ have studied Ag layers on Au substrates.
Here the Ag layer is the low plasma frequency cavity. With growing thickness again
and again a standing wave fits into the cavity and leads to resonances in an optical response, in this case the electroreflectance (ER) signal (see Sect. 3.6).
Figure 3.9 shows the oscillations of the ER signal due to the standing plasma waves.
They have been evaluated to derive the plasmon dispersion in silver /3.35/. The period of oscillation is the plasma wave length, not half that wavelength, in the experiment as well as in the calculation with the boundary conditions (2.52).

Next we want to shead some light on the general statements about the existence
conditions and the dispersion of the surface modes in certain limits and their relevance to experiments. FEIBELMAN has proven by an electron gas RPA-calculation
/3.36/ that on a metal surface with a smooth electron density profile there is always a surface plasmon with

$$\lim_{k \to 0} \omega_s(k) = \omega_p/\sqrt{2} \qquad (3.55)$$

irrespective of the shape of the surface density profile. Here ω_p is related to the
density of the bulk metal. Because many body electron gas calculations are usually
electrostatic approximations, this statement agrees with that of BOARDMAN /3.29 , 30/,

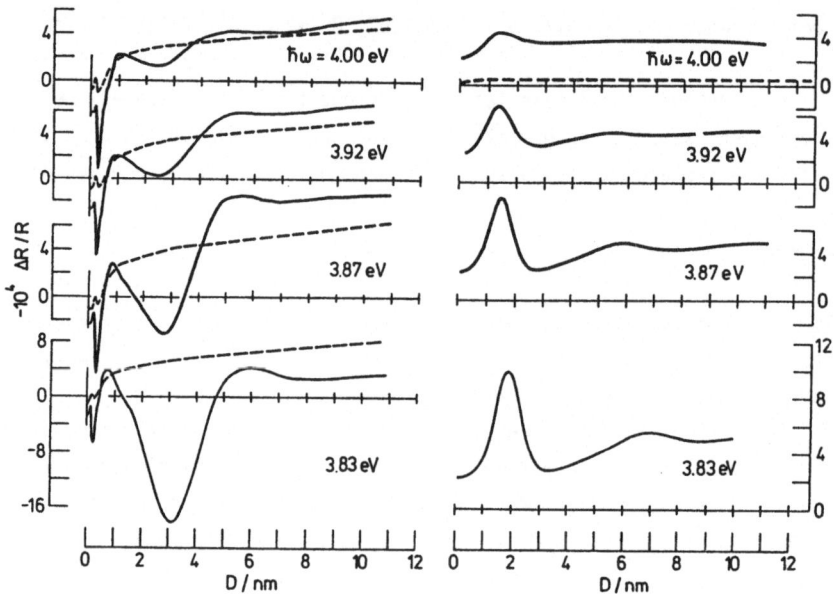

Fig. 3.9. Electroreflectance $\Delta R/R$ for Ag layers on Au(111) as a function of layer thickness D for p (————) and s (-----) polarization. Experimental curves on the *left*, calculated on the *right*/3.35/

that in the electrostatic approximation also the hydrodynamic model gives a mode which behaves like (3.55) irrespective of the surface density shape and that even including retardation always yields one mode "which appears to emanate from $\omega_p/\sqrt{2}$" /3.13/.

These statements contradict the plausible expectation, that if the decay of the surface density is, for instance, deliberately modelled by depositing a low density metal (Na) on a high density metal (Al), for large thickness of Na the eigenmodes should have frequencies $\omega_{Na}/\sqrt{2}$ for the mode centered at the very surface and $\omega^2 = (\omega_{Al}^2 + \omega_{Na}^2)/2$ at the interface from (3.52). So there is no mode at $\omega_{Al}/\sqrt{2}$ expected contrary to the theorem. In this argument, the second metal is only used to change the surface density profile at the surface of the bulk aluminum.

We will show that the expectation is not wrong. We can derive the eigenmode dispersion for a two step density decay at the surface (Fig. 3.3) from the zeros of the denominator (3.16) of the reflection coefficient. In the electrostatic limit $c \rightarrow \infty$ to lowest order in $k = (k_x^2 + k_y^2)^{1/2}$ we get (dropping the background polarizabilities)

$$\mathrm{DEN} \approx k^2 \cdot 4\varepsilon_1 e^{-\eta d}(\varepsilon_2 + 1)\left\{\frac{\varepsilon_1}{\varepsilon_2}\frac{\varepsilon_2 - 1}{\varepsilon_1 - 1}\cosh(\eta d) + \frac{\eta}{\mu}\sinh(\eta d)\right\} = 0 \quad . \tag{3.56}$$

For $k \neq 0$ only the zeros emanating from the zeros of the last two factors persist.

Therefore the conclusion is, that there is a surface mode for (neglecting damping)

$$\varepsilon_2 + 1 = 0 \qquad \omega = \omega_{p,bulk}/\sqrt{2} \tag{3.57}$$

in agreement with Feibelman's conclusion, and for

$$tgh(\eta d) = -\frac{\mu}{\eta}\frac{\varepsilon_1(\varepsilon_2 - 1)}{\varepsilon_2(\varepsilon_1 - 1)} \tag{3.58}$$

which yields the electrostatic $k \to 0$ limit for the standing wave surface modes. There is no solution for $\varepsilon_1 < 0$, $\varepsilon_2 < 0$, because $Re\{\eta,\mu\} > 0$. The standing wave modes appear only at frequencies, where plasma waves can propagate in the low density region. For $0 < \varepsilon_1 < 1$, $\eta = -i\eta'$ with $\eta' > 0$ and (3.58) changes to

$$tg(\eta'd) = \frac{\varepsilon_1}{\varepsilon_1 - 1}\frac{\varepsilon_2 - 1}{\varepsilon_2}\frac{\mu}{\eta'} \quad . \tag{3.59}$$

The first factor on the right is negative, while the others are positive, therefore $\pi/2 < \eta'd < \pi$ is necessary for the existence of the first higher mode. The solution $\varepsilon_1 = 0$, $\eta' = 0$ does not persist for $k \neq 0$ as mentioned above. The next mode appears for $3\pi/2 < \eta'd < 2\pi$. $\pi/2 < \eta'd$ means, that the surface needs to be soft enough to bind the first standing wave. This discussion is completely analogous to the one given by BOARDMAN et al. /3.13,30/. Relation (3.59) is slightly different from equation (201) in /3.13/ because we have used different boundary conditions, but qualitatively the results are the same.

The limit $k \to 0$ in the electrostatic approximation leads to fields according to (3.43 , 44), which do not at all decay away from the surface or interface. Averaging then over the involved density will always produce the bulk density, that's why the mode (3.57) appears in this limit.

A more relevant limit is reached by suppressing the terms proportional to $exp(-kd)$ in (3.16) for large d and analyse the remaining term in the limit $k \to 0$:

$$DEN \approx k^2 e^{-\eta d}(\varepsilon_1 + 1)(\varepsilon_2 + \varepsilon_1)\left\{\frac{\varepsilon_1}{\varepsilon_2}\frac{\varepsilon_2 - 1}{\varepsilon_1 - 1}\cosh(\eta d) + \frac{\eta}{\mu}\sinh(\eta d)\right\} = 0 \quad . \tag{3.60}$$

Here the third factor yields the surface plasmon (3.38) on the very surface of the overlayer, while the fourth factor gives the interface mode according to (3.52). The bracket is unchanged compared to (3.56) and so are the conditions for the higher standing wave modes in the surface layer. We therefore understand, that for larger d, the surface and the interface gradually decouple and each has its own mode. Only mathematically the very limit $k \to 0$ finally leads to nondecaying fields and dominance of the substrate density over any selvedge.

We see that the quantitative question arises, how small k is in relation to d and n_ℓ/n_{bulk}. The electrostatic $k \to 0$ limit is experimentally irrelevant anyway, because the low k dispersion is dictated by retardation. The dispersion curves are bent to lower frequencies, one branch down to zero, and there are no proper eigenmodes for $k < \omega/c$. But the interaction of light with the remnants of the standing wave eigenmodes can be found for $k < \omega/c$ as resonances.

We attempt to show in Fig. 3.10 the characteristics of the dispersion of the eigenmodes for the model of Fig. 3.3. The essentials are the same for other shapes of the surface density. For a free surface of a simple metal, the density decay is so steep, that the surface plasmon frequency is very much dominated by the bulk value $\omega_p/\sqrt{2}$, the (linear) increase of the frequency for larger k may be reduced to a plateau near $\omega_p/\sqrt{2}$ (Fig. 3.6 , 7), the small k dispersion is due to retardation (Fig. 3.8) and one can at best expect a single standing wave mode /3.32 , 34/.

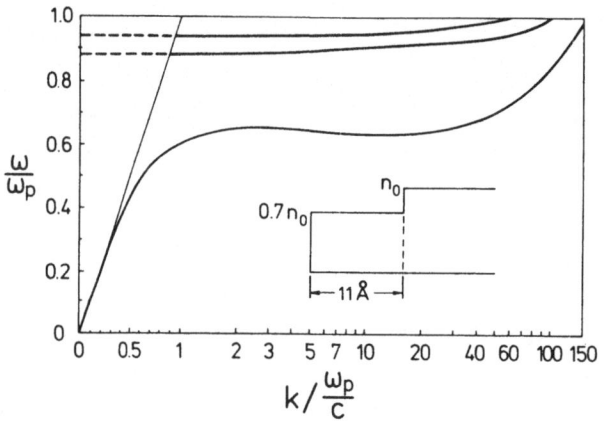

Fig. 3.10. Eigenmode dispersion of a stepped surface. The step has a width of 11 Å and a charge density of 0.7 n_{bulk}

According to (3.59) n standing wave eigenmodes can be expected, when $(2n-1)\pi/2 < \eta'd$ ($n = 1,2,3,...$) or $d > n\Lambda/2 - \Lambda/4$ with the plasmon wavelength Λ in the cavity region. Here we find a periodicity with $\Lambda/2$ for the eigenmodes existence condition. If the resonances (left of the light line) related to these eigenmodes are studied in an optical experiment, the periodicity (with layer thickness, for instance) need not be $\Lambda/2$. It has been found to be Λ in the reflection from thin layers (Sect. 3.2) and in the electroreflection from deposited layers (Fig. 3.9).

3.5 Optical Properties of Metal Layers on Metal Substrates

Systematic studies of the optical properties of metal layers on metal substrates have been carried out in the group of ABELES /2.30 ; 3.37 - 39/. Especially silver layers on Al and Au substrates have been investigated near the plasma frequency of silver. Here we present a few results of these publications which are good examples for the relevance of plasma waves in optics.

For a metal layer of given thickness on a substrate with given dielectric properties, the complex dielectric constant at fixed frequency can be derived from fitting the calculated reflection coefficient to the measured reflectivity at a number of different angles. This fitting can the done by reflectivity calculations according to standard optics and according to the nonlocal optics including plasma waves. The results are drastically different. Fig. 3.11 shows results of such a determination of ε for a layer of silver on an Al substrate /3.37/. The reflectivity was measured here in an especially sensitive attenuated total reflection (ATR) arrangement /3.18 , 19/.

The standard analysis (Fig. 3.11a) yields a lower and lower real part of ε for

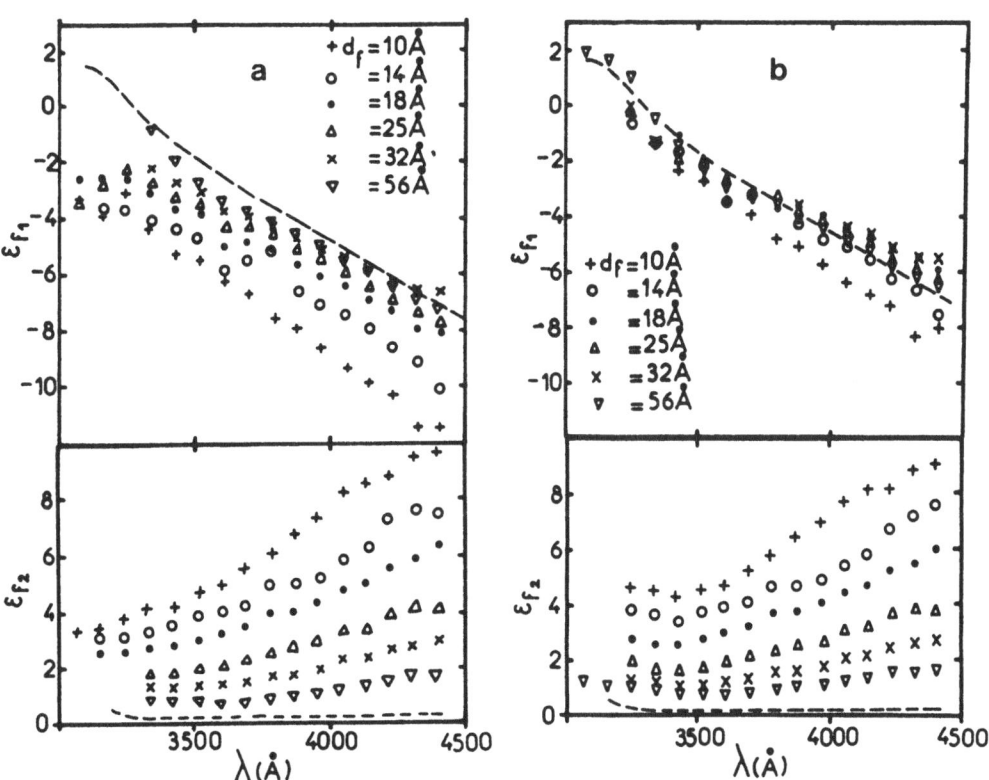

Fig. 3.11a,b. Real part ε_{f1} (*top*) and imaginary part ε_{f2} of the dielectric function of silver films of thickness d_f on an Al substrate. (a) The evaluation of ATR measurements using standard optics. (b) The evaluation including the plasma waves. (– – –) show the bulk silver values /3.37/

decreasing thickness of the silver layer. The impression is, that the high electron density of the Al substrate penetrates the silver up to 20 Å and increases ω_p or reduces $\varepsilon_{eff} = \varepsilon_b - \omega_p^2/\omega^2$ for the thinner layers. The same data analysed by nonlocal optics tell a completely different story (Fig. 3.11b). The bulk dielectric function represents the dielectric properties well down to very thin layers, at least for the real part of ε. The plasma wave optics couples properly the silver and the aluminum system so that the compound system shows enough influence of Al, even when

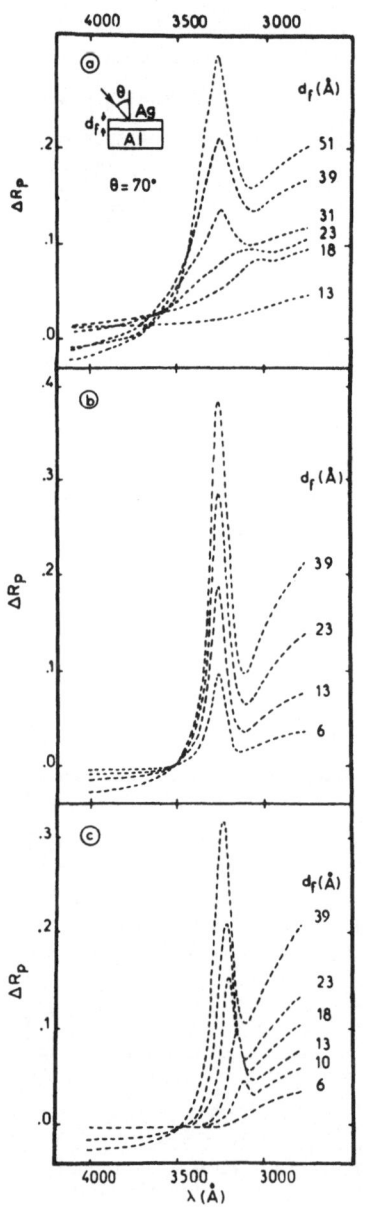

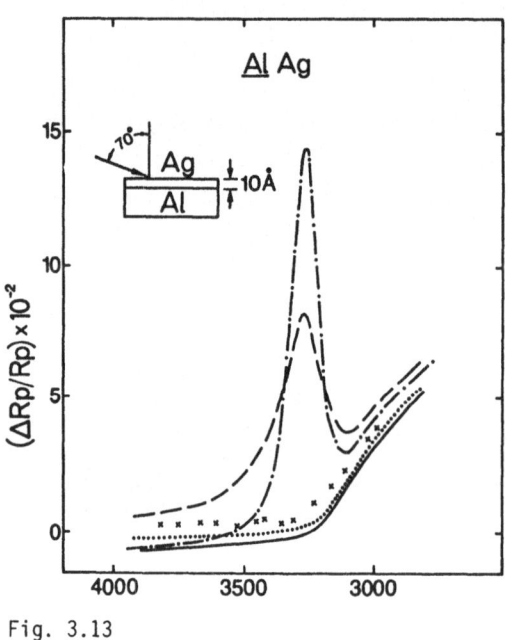

Fig. 3.13

◄Fig. 3.12a-c. The reflectance difference ΔR_p = R(free) - R(covered) for aluminum without and with a silver film of thickness d_f on the surface. The light was p polarized. (a) measurements, (b) computed R_p from standard optics, (c) computed ΔR_p including plasma waves /3.38/

Fig. 3.13. Relative reflectance difference 2[R(free) - R(covered)]/[R(free) + R(covered)] for Al covered by an Ag film of 10 Å. (× × ×) measurements, (—·—·—) standard optics, (— — — —) damping increased by factor 10, (———) nonlocal optics, (·····) damping increased by factor 10 /3.39/

the dielectric properties of the silver layer are unchanged. This interpretation is discussed more intensively in Sect. 3.6. It can also help to understand the next example.

Fig. 3.12a shows the difference in reflectivity of an Al surface without and with a silver layer on top /3.38/. Near the silver plasma frequency this difference in reflectance has a pronounced maximum, which shifts to higher frequency and slowly vanishes with decreasing film thickness. A calculation by standard optics shown in Fig. 3.12b gives a strong maximum down to the thinnest layers without any shift while the nonlocal calculation in Fig. 3.12c shows the essential features of the measurements, the shift to higher frequencies and the disappearance of the peak for smaller thicknesses. In Fig. 3.13 a more detailed comparison is shown for a layer of 10 Å thickness where the structure has disappeared in the measurement /3.39/. Even a drastic increase in damping by one order of magnitude does not wipe out the resonance peak predicted by standard optics, while the measurements do not show the peak and agree well with the nonlocal calculation, and this agreement does not depend on some uncertainty about the damping.

We again understand these results by saying that the nonlocal optics spreads charge perturbations and couples to the substrate in a way, that the compound system looses its resonance at the plasma frequency of silver for very thin layers, while in standard optics the confinement of the charge perturbations to a singular surface plane leaves the resonance properties unchanged down to the thinnest silver layers.

3.6 Electroreflectance Spectra at Silver Surfaces

Electroreflectance (ER) is an optical technique which can probe the electronic structure at metal surfaces. Reflectivity differences due to electron density changes induced by static electric fields at the surface are measured. In order to achieve detectable changes, the charge density variations must be large enough, i.e. the fields must be extremely high. Therefore the most efficient capacitor, a double layer at a metal-electrolyte interface, is employed /3.40/. It needs some familiarity with the method and its results to get convinced that this technique can be used to investigate the properties of metal surfaces. Disagreement between measurements and optical calculations which consider only the metal could raise the hope that the technique may yield some information about the electrolyte part of the interface, and a number of experiments is actually directed towards this aim. We show here two examples, where electroreflectance results which look incompatible with a standard optics calculation can be fully understood when nonlocal metal optics is employed.

The analysis of the electroreflectance measurement assumes (until disproved) that only changes of the metal properties induce the reflectance modulations. These changes are calculated by the concept proposed by McINTYRE /3.41/: In a surface region of an estimated thickness of a few Angström the conduction electron density is changed and therefore this region changes its dielectric properties according to (see Sect. 2.7):

$$\varepsilon(\omega) = \varepsilon_b(\omega) - \omega_p^2(n_0 + \Delta n)/(n_0\omega^2) \quad . \tag{3.61}$$

The change of the surface charge due to a variation of the voltage at the electrolytical cell can be calculated from a knowledge of the double layer capacitance. The change in electron *density* Δn is then related to the thickness of the modulated layer. Because a small variation in a thick region has an effect comparable to that of a large variation in a thin region, the estimate of the layer thickness does not very critically affect the results. In the calculations presented here, the chosen thickness was 4 Å in most cases (taken from a fit to the surface plasmon dispersion shown in Sect. 3.3); in one example it was 1 Å. The first example is an electroreflectance spectrum from a massive Ag single crystal (100) surface /2.29/.

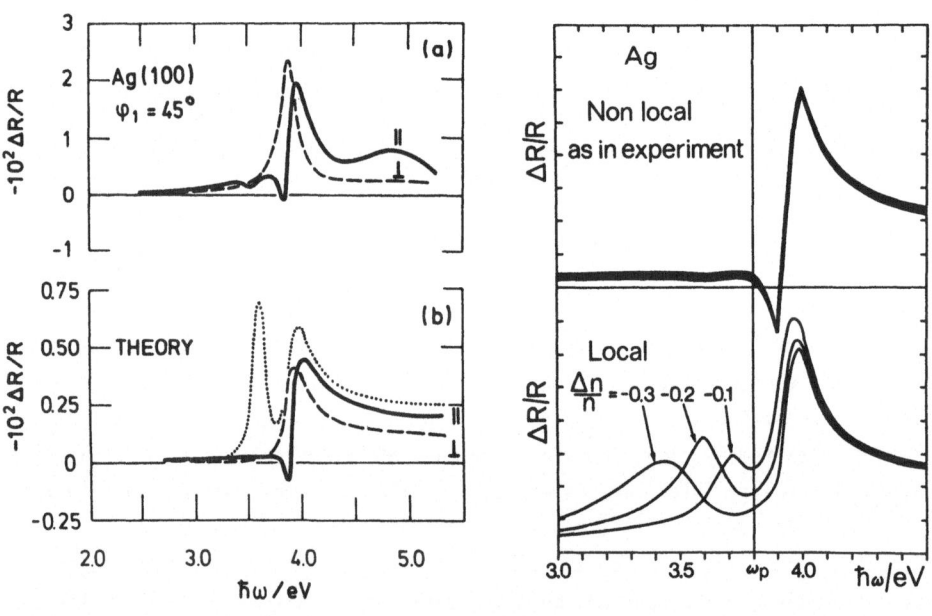

Fig. 3.14   Fig. 3.15

Fig. 3.14a,b. Electroreflectance spectra of Ag at non-normal incidence ($\varphi = 45^0$). (a) Experimental results for an Ag(100) electrode with p (———) and s (— — —) polarized light. (b) Calculated spectra: (— — —) s polarized, (———) p polarized with plasma waves, (·····) p polarized with local optics. The relative conduction electron density in a 1 Å thick surface layer is reduced from 1.0 to 0.8 /2.29/

Fig. 3.15. Calculated electroreflectance spectra for thick silver for various $\Delta n/n$ values. Surface layer thickness 4 Å

Figure 3.14a shows the relative reflectance difference $\Delta R/R$ for s- as well as for p-polarized light. Near the plasma frequency of silver both curves show a resonance structure. For s-polarized light the plasma waves play no role (see Sect. 3.1) and the assumptions about the optical properties and the parameters of the model can be tested by finding a single resonance peak at the right position as shown by the dashed line in Fig. 3.14b. Now all parameters are fixed and one tries the calculation for p polarization. Standard optics yields the dotted curve completely off the measurements. The inclusion of the plasma waves into the optical calculation yields a result in very good agreement with the experiment.

Another characteristic is shown in Fig. 3.15 /3.42/. When the surface charge deficit which causes the reflectance change is increased, the spectrum of $\Delta R/R$ does not change its form as also predicted by the nonlocal calculation. (Only the amplitude of the signal is proportional to $\Delta n/n$). In the calculation with standard optics the second peak also seen in Fig. 3.14b (for slightly different parameters) moves to lower energies. It actually shows up at the plasma frequency of the surface layer with the reduced charge density.

A second example for an improved analysis of electroreflectance spectra by nonlocal optics concerns a study of silver layers on a copper substrate /3.42 , 43/.

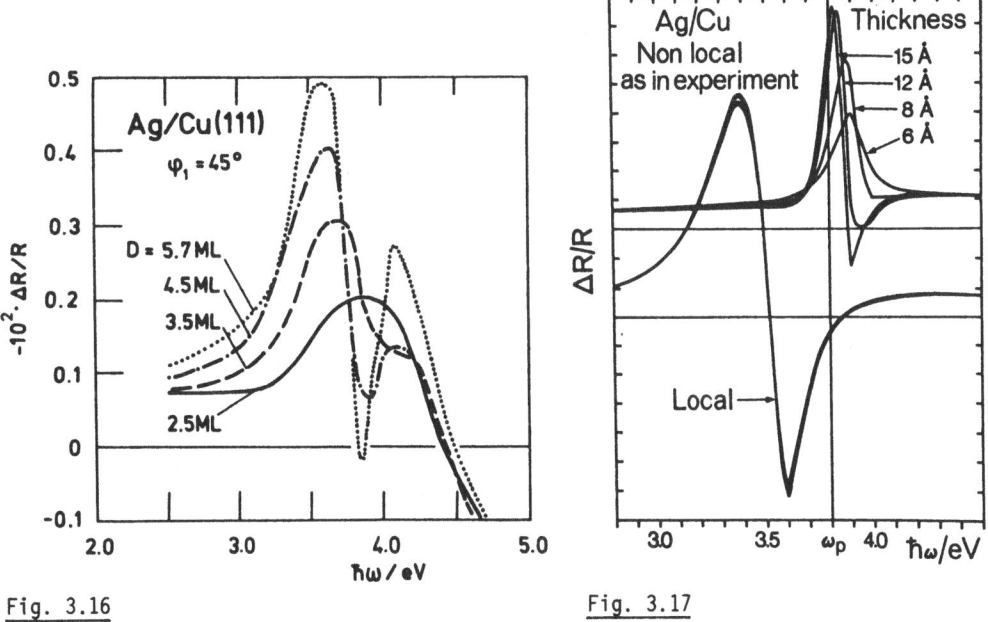

Fig. 3.16

Fig. 3.17

Fig. 3.16. Measured ER spectra for Ag layers of various thicknesses on a Cu substrate. p polarization, $\varphi = 45^\circ$, ML = monolayer /3.42/

Fig. 3.17. Calculated ER spectra for Ag layers of various thicknesses on a Cu substrate. p polarization, $\varphi = 45^\circ$

Figure 3.16 shows $\Delta R/R$ spectra for very thin silver films of a few monolayers (ML). The layers were grown by electrodeposition with an accurate control of layer thickness. Remember that ΔR is induced here by changing the surface charge in the silver layer, while in the previous Sect. 3.5 ΔR was taken between the systems without and with the silver layer. The ER spectra change with increasing thickness from a single maximum to a higher maximum at reduced frequency followed by a deep minimum. The same characteristic behaviour is seen in the nonlocal calculation in Fig. 3.17 while the standard optics calculation does not show any sensitivity to the thickness of the deposited silver layer.

Both examples show again, that in standard optics with divergence free fields inside the metal, the very surface plane with its screening properties dominates the picture; in Fig. 3.17 this has the consequence that it does not matter how much silver is between the surface plane and the bulk copper substrate. In Fig. 3.14 , 15 the surface plane shows its own plasma resonance when the reflectivity is calculated with transverse waves only. Nonlocal optics means keeping charge densities finite by speading the charge perturbations. Then these perturbations recognize variations of n_0 near the surface; they couple to the charge density inside. A single compound resonating system is formed with only one resonance in Fig. 3.14 , 15 and a sensitivity to the silver layer thickness in Fig. 3.17. The minimum after the maximum in Fig. 3.16 , 17 is connected with a standing plasma wave with $\lambda \approx 30$ Å in the silver layer of 15 Å. These standing wave resonances have been further investigated by electroreflectance /3.35/. Results are shown in Fig. 3.9 and discussed in Sect. 3.4.

It does not seem reasonable to present here formulas for the evaluation of electroreflectance measurements. The methods outlined in Chap. 2 can be straightforwardly applied, when a model containing a sequence of homogeneous layers is specified. At every interface the boundary conditions (2.31 , 45 , 52) yield three or four linear equations and this system of equations must be solved. We found it more flexible to solve the full system on a computer than to consider expansions for small layer thickness, which might be possible occasionally.

3.7 Ellipsometry from Metal Surfaces

In ellipsometry the change of polarization accompanying the reflection of linearly polarized light is determined. The ratio of the reflection amplitudes for p- and s-polarized light is written as /3.44 , 45/

$$R_p/R_s = \exp(i\Delta) \cdot \mathrm{tg}\,\Psi \quad . \tag{3.62}$$

Ψ and Δ are the ellipsometry parameters containing the information about the dielectric properties of the reflecting system. Especially, changes $\delta\Psi$ and $\delta\Delta$ due to changes of the surface properties can be measured with very high accuracy /3.46/.

When ellipsometry is carried out near the plasma frequency of the investigated metal nonlocal optics changes the analysis. In the case of reflection from a metal halfspace, the ratio (Eq.(3.6)/Eq.(3.31)) has to be expressed by phase and modulus. An article of ABELES and LOPEZ-RIOS /3.47/ presents the relevant formulas. Because the reflection amplitude for a thick piece of metal is only slightly changed by spatial dispersion as discussed in Sect. 3.1, also the effect of spatial dispersion on Ψ and Δ is very small in this simple case. Measurable effects develop when some metal layer of thickness smaller than a few times the plasma wave decay length η^{-1} is involved.

Such an example was studied in the case of oxygen adsorption on a silver substrate /3.48/. The differences $\delta\Psi$ and $\delta\Delta$ which are caused by the oxygen adsorption are evaluated to yield a real and imaginary part of an effective dielectric constant for the adsorbed layer. At a fixed frequency $\varepsilon(\omega) = \varepsilon_r + i\varepsilon_i$ can be fitted to $\delta\Psi$ and $\delta\Delta$. This fitting procedure has usually two solutions /3.44 , 49/. In the case of oxygen on silver both solutions are implausible. One changes the sign of the imaginary part ε_i near ω_p, the other has a negative real part ε_r in the whole frequency range studied /3.46/. Fitting by standard optics as well as by nonlocal optics yields nearly the same unsatisfactory results.

Improvement was achieved in this case by a more sophisticated model, not really by additional parameters. As discussed in several sections of this article, some optical effects are sensitive to the smooth decay of the charge density at the surface of the unperturbed metal. For the optical treatment described in Chap. 2 such a continuous decay in density can be approximated by a selvedge layer of small thickness and decreased density.

In /3.48/ the proposition was tested, that such a surface layer of decreased electron density on the free silver surface is transformed under oxygen adsorption into an effective oxid layer with ε_{eff}. Such a layered system can be reasonably treated only by nonlocal optics, otherwise separate resonances will appear, as also shown in /3.48/. By fitting ε_{eff} of the oxid layer to the measurements within this more elaborate model a positive real part and an imaginary part of definite sign was evaluated.

It is not claimed, that this analysis is the last word about the dielectric properties of adsorbed oxygen layers. This example only shows that also in ellipsometry there can be cases, where progress in the data analysis needs the nonlocal metal optics.

Another recent case is the experiment of CHAO and COSTA /3.50/ who measured the dependence of Ψ and Δ on the charge at a gold surface. They concluded from a standard optics analysis, that the penetration depth for static external fields increased linearly with the (positive) charge on the surface. This result is not in agreement with microscopic calculations and with the experience in the case of electroreflectance (Sect. 3.6). KEMPA /3.51/ has shown, that analysing CHAO and COSTA's measurements with nonlocal optics yields an essentially constant penetration depth of about 2.5 Å.

Nonlocal effects in ellipsometry in relation to experiments on tungsten are discussed in /3.52/.

3.8 Resonances in Small Metal Spheres

That colloidal particles of gold in glas have a deep red colour obviously different from the gold metal which is yellow in reflection and green in transmission was known already by the artists of the medieval cathedral windows. The optical properties of small metal particles are different from the bulk metal for several reasons:

a) a small particle (diameter smaller than the wavelength of light) shows a characteristic scattering and absorption resonance, the MIE resonance /3.53/ which is not found with the bulk metal;

b) the size of the spheres can decrease below the bulk mean free path of the electrons yielding a general increase of the optical absorption due to surface scattering of the electrons;

c) at very small diameters the electron energy spectrum should change from the quasi continuum in the metal to a discrete spectrum as found in extreme in a single atom; this quantum size effect is expected to change the optical properties at diameters around 10 Å and smaller;

d) further changes are due to the interaction between suspended particles and between the particles and the matrix which is usually used to keep the particles suspended.

Recently several studies of the optical resonances in thin metal spheres were published in relation to experiments on optical absorption /3.54 - 59/ or photoemission measurements of small metal particles in aerosols /3.60/ or embedded in matrices. This topic is an extended field on its own and cannot be reviewed carefully in this context. We refer the reader to the special review article of PERENBOOM et al. /3.61/ for instance or to the articles of RUPPIN and LUSHNIKOV et al. in /3.13/. Here we will only outline the main arguments and possibly sort out some irritations about the increase or decrease of the resonance freqeuncy with decreasing particle size. We will be concerned here with the results for single spheres. In relation to point (d) we only mention a few references /3.61 - 66/.

For a single small metal particle it is most important to understand the predictions of macroscopic optics, because these also may dominate the results of quantum mechanical calculations and obscure a real quantum size effect. When a sphere with dielectric constant ε_{sph} is in a homogeneous electric field E_0 in a medium with ε_M, the field inside is /3.67/

$$E_i = \frac{3\varepsilon_M}{\varepsilon_{sph} + 2\varepsilon_M} E_0 \quad . \tag{3.63}$$

Therefore the field inside increases like resonant when

$$\varepsilon_{sph} + 2\varepsilon_M = 0 \quad . \tag{3.64}$$

For a simple metal sphere

$$1 - \frac{\omega_p^2}{\omega^2} + 2\varepsilon_M = 0 \tag{3.65}$$

$$\omega_{res}^2 = \frac{\omega_p^2}{1 + 2\varepsilon_M} \quad , \tag{3.66}$$

in vacuum: $\quad \omega_{res} = \omega_p / \sqrt{3} \quad . \tag{3.67}$

This is the MIE-resonance /3.53/, which was derived for spheres small compared to the wave length of light, i.e. effectively in a homogeneous field. The reason for the large fields is the resonant oscillation of the whole sphere of electrons versus the sphere of ion cores. (A thin metal layer oscillates with ω_p, the eigenfrequency of the bulk metal, normal to its surface.) On one side of the sphere a negative charge appears, on the other side a positive charge, changing periodically in time. Standard optics confines these charges to surface charges. In this approximation, the resonance frequency does not depend on the size of the sphere provided it is smaller than the light wave length. For larger diameters, there is a slow decrease of the eigenfrequency due to retardation /3.68/.

From the previous sections it is obvious, that changes are expected, when the charge density near the surface is treated more carefully. A calculation in the hydrodynamic approximation shows again the essential effects /3.69, 70/.

Because the electron gas develops an additional restoring force when charge accumulation and depletion are near to each other (the gradient in the electron gas pressure is responsible for spatial dispersion, Sect. 2.2) the resonance frequency increases for decreasing sphere radius, when spatial dispersion is taken into account (Fig. 3.18). This is analogous to the increase of the surface plasmon frequency for small wavelengths (Sect. 3.3). One may interpret the charge perturbation on the sphere as a surface plasmon around the circumference of the sphere. A smaller size leads to a shorter wave length and a higher frequency. As for a planar surface this result is obtained when the model for the electron distribution is a sharp step ($\lambda = 0$ in

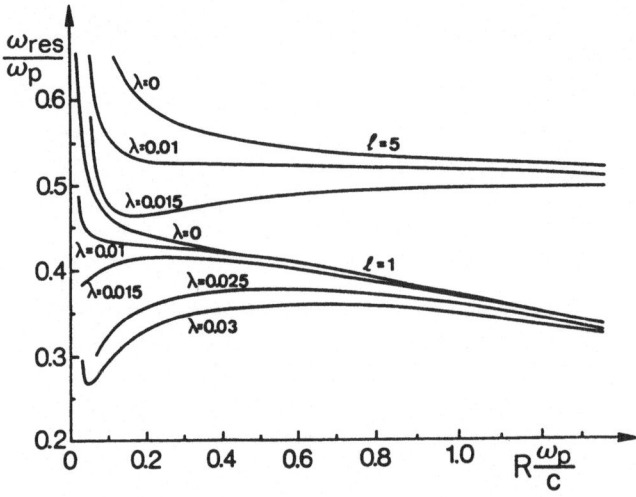

Fig. 3.18. Resonance frequency ω_{res} ($\ell = 1$ and $\ell = 5$) versus sphere radius R for a double step surface in the hydrodynamic approximation. Parameters for silver in glass. Increasing λ means increasing diffuseness of the surface /3.69/

Fig. 3.18). The same result was derived by more lengthy RPA calculations for the same model /3.71 - 74/.

On the other hand several papers have been published, which derive a decrease of the resonance frequency when the sphere gets smaller /3.69 , 75 - 77/. For the hydrodynamic calculation it is obvious (Fig. 3.18, $\lambda \neq 0$), that the soft density profile at the metal surface causes a decrease of the frequency as discussed for the planar surface in Sect. 3.3. When the sphere size decreases there is less and less volume with the full bulk charge density; a lower average charge density determines a lower plasma frequency. The HD calculation with a two step surface density demonstrates this behaviour in Fig. 3.18 /3.69 , 78/. Therefore for decreasing sphere radius, the two effects are competing: Increase of ω_{res} due to the electron gas pressure and decrease due to the lowering of the effective charge density in connection with a soft gradient at the surface. As on the plane surface the result of this competition depends on the system. Roughening of the surface will yield a softer profile and favour a decrease of ω_{res}. These conclusions all can be read off from Fig. 3.18. These considerations render it plausible, that well shaped particles formed in the gas phase and matrix isolated afterwards show a blue shift of the absorption frequency for smaller seizes /3.59/, while crystallites in stained glass /3.57 , 58/, with probably rough surfaces show a red shift. Also in the classical Mie-theory without spatial dispersion, a continuous density profile at the surface leads to a decrease of the resonance frequency /3.79/ with decreasing diameter.

So far the influence of a decreasing diameter on the *electron states* has not come into play. The mean free path is soon larger than the particle diameter. GENZEL et al.

/3.55/ have included the additional surface scattering in an increased damping

$$\gamma = \frac{1}{\tau} = \frac{1}{\tau_{bulk}} + \frac{1}{\tau_{surface}} = \frac{1}{\tau_{bulk}} + \frac{v_F}{R} \qquad (3.68)$$

in a Drude dielectric function. Such an approximation can be justified from quantum mechanics /3.80/. A decrease of the resonance frequency results when the particles get smaller (like for a stronger damped pendulum). The same authors discussed also the influence of a discretisation of the electron spectrum on the plasma frequency. They derived an increase which more than compensates the slowing down by the surface scattering.

When these corrections on $\varepsilon(\omega)$ are introduced, one gets changes of the resonance frequency already without considering plasma wave effects, but this is probably a poor approximation since over the range of radii from 100 Å to 10 Å the plasma effects appearently dominate the picture and should be included first. A consistent microscopic calculation comparable to Feibelman's for the planar surface (3.84 - 86, Sect. 3.9) naturally includes both aspects /3.98/.

Up to now we have discussed only the dipolar mode of excitation of the sphere which is the only one relevant in optical excitation as long as the diameter of the sphere is smaller than the wave length /3.81/. There are higher angular momentum eigenmodes of metal spheres with an angular dependence of the fields proportional to the spherical harmonics $Y_\ell^m(\vartheta,\varphi)$ /3.82/. An argument like the one leading to (3.66) yields for the higher order resonant frequencies /3.69 , 81/

$$\omega_\ell^2 = \omega_p^2[1 + \varepsilon_M(\ell+1)/\ell]^{-1} \quad . \qquad (3.69)$$

The dipole mode (3.66) is obtained for $\ell = 1$, while for large ℓ the eigenmodes approach the frequency of the surface plasmon for a plane (3.38) because for shorter wave lengths the strong confinement of the fields to the surface makes the curvature less and less important.

These resonance frequencies shift with decreasing size of the sphere to higher values for a sharp surface and to lower values if the soft electron density gradient is taken into account (Fig. 3.18) in the same way as explained for the dipole mode. The higher modes are more sensitive to the diffuseness of the surface charge distribution because the perturbation is more confined to the surface, but at very small sizes the short wavelength effect due to the electron gas pressure, which increases the frequency, again takes over (Fig. 3.18).

These higher ℓ modes are important for the electron energy loss at small metal particles, as discussed in /3.81/. High energy electrons passing through metal spheres excite all these modes with a certain probability. For small spheres (50 Å) the excitation of the dipole mode $\ell = 1$ dominates. For larger spheres higher ℓ modes become more and more probable because the exciting field is not homogeneous any more and the energy loss maximum shifts towards the value $\omega_s = \omega_p/\sqrt{1+\varepsilon_M}$, which it should reach for a planar surface.

Many investigations of the resonance frequency and the field distribution near small particles are published in studies of surface enhanced Raman spectroscopy (SERS), because the large fields connected with the optical excitation of the resonance are part of the enhancement mechanism.

3.9 The Electric Fields Near the Surface Inside the Metal

In Sect. 2.3 we have formulated one aim of the approximate surface treatment via boundary conditions: the calculation of the proper *asymptotic* fields. It is more ambitious to demand also a useful representation of the fields near the surface.

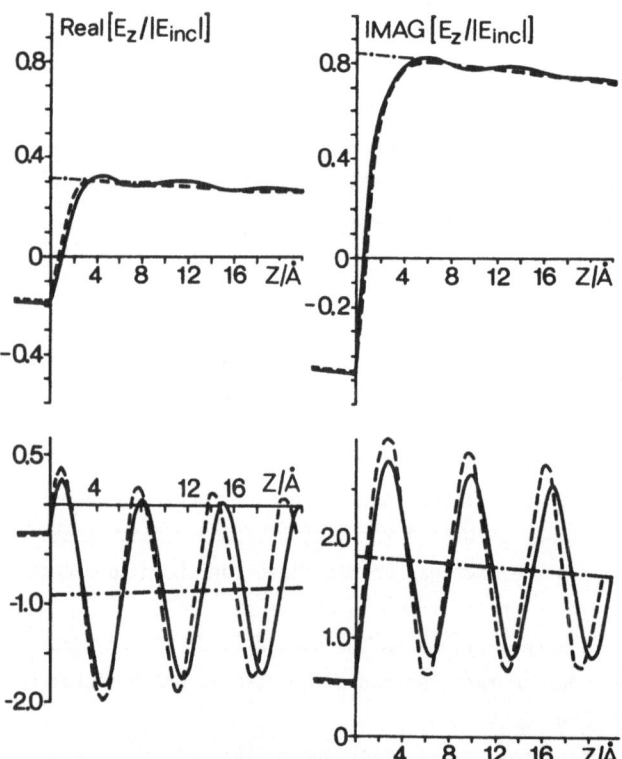

Fig. 3.19. The variation of the electric field E_z normal to the surface. The fields are normalized by the modulus of the incoming wave. (———) are from KLIEWER /3.83/, (———) is from the hydrodynamic approximation, (—·—·—) are the fields from standard optics. $\gamma/\omega_p = 10^{-2}$ *top*: $\omega = 0.8\,\omega_p$, *bottom*: $\omega = 1.2\,\omega_p$. From /3.87/

It has been claimed several times /3.83 , 84/ that the hydrodynamic approximation as presented here behaves poorly in this respect. We will show in this and the next section, that one can also justify the point of view that our HD approximation does surprisingly well. This dispute can be judged by comparing with fields calculated via more sophisticated treatments of a metal surface, or, as done in the next section, to an experiment which tests the field near the surface.

Any microscopic model of a metal surface on the level of quantum mechanics implies finite charge densities and therefore continuous electric fields. This is one deviation from classical optics common to microscopic surface models and also essential for the hydrodynamic nonlocal' optics. For frequencies below ω_p the nonlocal field essentially connects continuously the classical field value outside to the classical value inside (see Fig. 3.19) with an exponential decay length η^{-1} from (2.27). The quantum mechanical models produce in addition Friedel oscillations as around any abrupt perturbation, which decay $\sim z^{-2}$ with a typical wavevector $k = (2m\omega/\hbar + k_F^2)^{1/2} - k_F$ (see Sect. 4.5). These oscillations have small amplitudes as seen in Fig. 3.19 , 20. They derive from a more complicated response function of the electron gas than the 'plasmon pole approximation' (2.19). This small difference

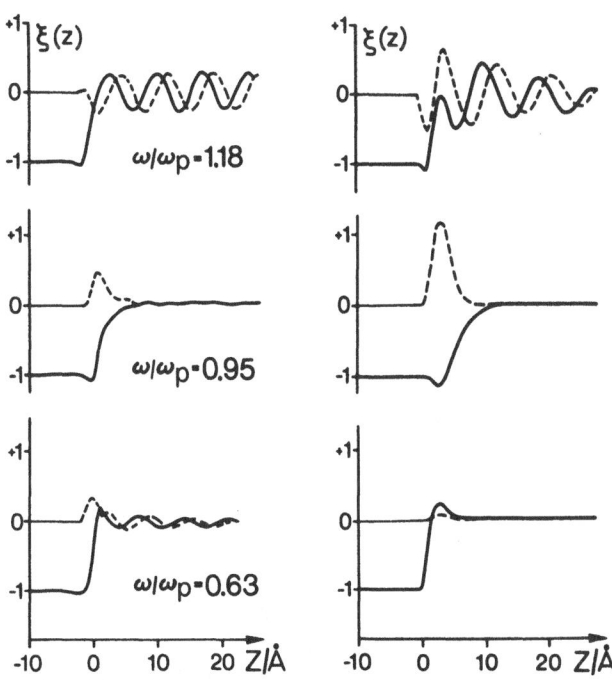

Fig. 3.20. Normalized electric field near the surface. $\xi(z) = [E_z(z) - E_z(\text{inside})]/[E_z(\text{inside}) - E_z(\text{outside})]$. (———) $\text{Re}\{\xi(z)\}$. (— — —) $\text{Im}\{\xi(z)\}$. *Left side:* according to FEIBELMAN /3.84 , 86/. *Right side:* hydrodynamic approximation with surface step and homogeneous damping /3.87/

between the hydrodynamic model and a random phase approximation is often described by saying that the hydrodynamic approximation (HD) does not take into account single particle excitations. This is misleading, because in certain sophisticated treatments such as RPA /3.84/ any response is via single particle excitations, also the plasmon. The HD treats those coherent single particle excitations which form a collective mode. Neglected are incoherent single particle excitations which interfere destructively in the charge density perturbation. The energy loss which goes into these single particle excitations and which because of incoherence cannot come back to field energy is effectively included in the damping parameter γ in (2.15). The ground state charge density ρ_0 as well as the perturbation $\delta\rho$ is assumed smooth (often monotonous), while its representation as a sum of a finite number of wavefunctions with definite phases at the surface necessarily leads to small Friedel oscillations.

For frequencies above ω_p the plasma wave is fully developed and dominates the picture in any treatment (Fig. 3.19 , 20) even the phase is well given by the HD. In Fig. 3.19 the fields from the HD are compared to the fields published by KLIEWER /3.83/, whose method of treating the surface response is discussed in Sect. 4.3. He employs the full RPA dielectric function of an electron gas.

A more detailed study for a surface with a soft density gradient was published by FEIBELMAN /3.84 - 86/. These results were compared in /3.84/ with results from HD for a sharp density step and the HD method was disqualified by disagreement. A fair comparison is shown in Fig. 3.20 /3.87/ where also in the HD optics our standard simulation of a soft surface by a single step of smaller density according to Fig. 3.3 has been employed. Plotted are the real and imaginary parts of $E_z(z)$ normalized by the field values given by standard optics, which are the limits in the homogeneous regions. It is easy to make the agreement much closer when the HD parameters are fitted to this problem, which was not done in /3.87/. Figure 3.20 shows that nearly all essential features are produced, especially the peak near the surface below ω_p. This peak is due to a plasma wave starting in the low density region and decaying exponentially in the high density region, if $\omega < \omega_p$ (bulk). As discussed in Sect. 3.4 the plasma wave has to resonate in the "cavity" of low surface electron density in order to yield the large peak in Fig. 3.20. This is the proper interpretation and not a large single particle contribution due to singular fields /3.83, 88/. The fact that this starting plasma wave is also produced by a sophisticated RPA treatment tells us that our two step model, which was fitted to the surface plasmon dispersion (Sect. 3.3) is transferable even with respect to this detail. For a large peak the surface density profile must be soft enough in order to reach the first standing wave resonance (Sect. 3.4). Also, in a paper by FEIBELMAN /3.85/ it can be seen that the surface peak in Fig. 3.20 drops drastically, when the surface density gradient becomes steeper.

3.10 The Photoemission Yield from Metal Surfaces

Nonlocal metal optics has been most extensively discussed in connection with the photoemission yield. Photoelectrons with energies around 15 eV above the Fermi level have a very short escape depth, about 10 Å /3.89/. They therefore should probe the electromagnetic field near the surface. This was pointed out especially by KLIEWER /2.25/.

There are three other effects, which enhance the influence of the field near the surface in the photoemission process. In a homogeneous free electron gas there is no possibility for photoexcitation by slowly varying fields, e.g. the long wave length transverse fields, because all matrix elements $<\Psi_k \nabla \Psi_{k'}>$ vanish. Only the potential gradient at the surface and consequently the deviation of the wave functions from simple sin-functions overcomes this selection rule for the electron gas. Therefore the contributions to the photoemission matrix element will heavily weigh the electric field near the surface. In nearly free electron metals like aluminum this dominant surface emission has only a weak bulk background due to the crystal potential or band structure.

In addition the phototransitions in the electron gas become allowed, if the electric field varies strongly in space and cannot be taken out of the matrix elements. This is the case again near the surface as discussed in the previous Sect. 3.9. Finally, the operator in the phototransition matrix element is $(\mathbf{A} \cdot \nabla + \nabla \cdot \mathbf{A}) = 2\mathbf{A} \cdot \nabla + \text{div } \mathbf{A}$. Usually transverse fields are considered and the last term is dropped. Near the surface where the plasma waves contribute to the fields, div $\mathbf{A}$ is nonzero. It even can yield the dominant contribution near the surface in an electron gas model /2.25 , 3.84 , 90 - 93/.

The photoemission yield from a metal surface has been calculated in several approximations. The results are discussed in the following paragraph. They are plotted in Fig. 3.21b-d /3.87/ and are compared to the experimental yield from aluminum /3.90 , 94/ in Fig. 3.21a. Every model with continuous fields at the surface yields very small surface fields inside the metal at the plasma frequency (see Sect. 3.1) leading to the sharp minimum for the photoyield at ω_p. Only standard optics (Fig. 3.21, curve 7) does not show this minimum.

The concept of assuming the photoemission yield proportional to the power absorption was taken up in this context by KLIEWER /2.25/. He multiplied the electromagnetic energy absorption density a(z) by an exponential escape probability exp(-z/ℓ) with the escape depth ℓ and integrated over z to get the total photoemission yield:

$$Y(\omega) \sim \int_0^\infty a(z,\omega)e^{-z/\ell}\, dz \quad . \tag{3.70}$$

KLIEWER and later APELL /3.88/ took a(z,ω) = Re{$\mathbf{E}^* \cdot \mathbf{j}$} and got results shown in Fig. 3.21c in severe disagreement with experiment. Above ω_p the integral over Re{$\mathbf{E}^* \cdot \mathbf{j}$} is

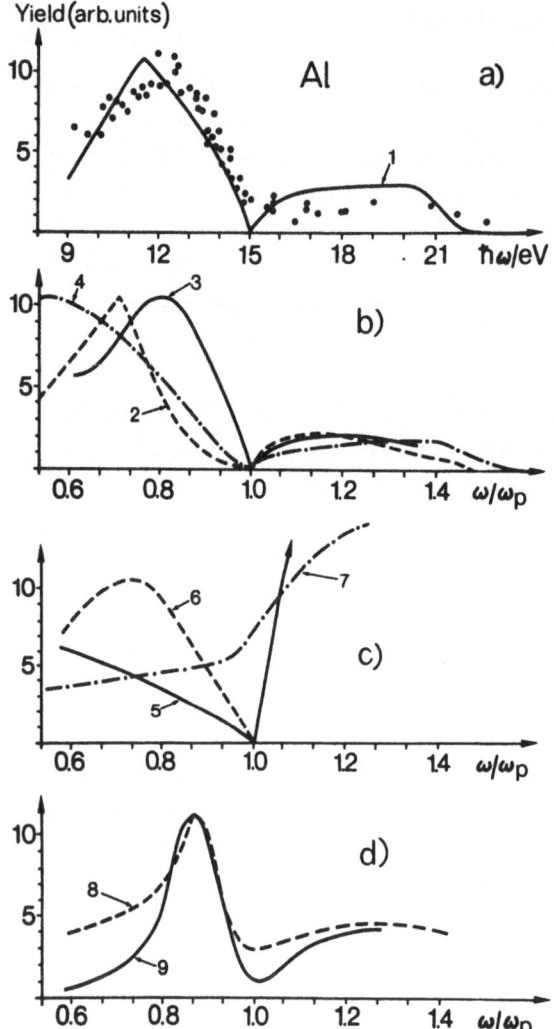

Yield (arb. units)

Al   a)

$\hbar\omega/eV$

b)

ω/ω_p

c)

ω/ω_p

d)

ω/ω_p

Fig. 3.21a-d. The photoelectron emission yield from a metal surface at frequencies near ω_p.
(**a**) Measured yield: *1* from /3.94/ and (•••••) from /3.90/.
(**b**) Calculations via matrix elements: *2* ENDRIZ /3.92/, *3* FEIBELMAN /3.90/, *4* BARBERAN et al. /3.93/.
(**c**) Re{$\mathbf{E}^*\cdot\mathbf{j}$}: *5* KLIEWER /2.25/, *6* APELL /3.88/, *7* standard optics,
(**d**) HD optics with a surface step: *8* γj^2, *9* $|\int E_z(z)dz|^2$ /3.87/

more than an order of magnitude larger than the measured yield, if we normalize at the maximum below ω_p. It was explained previously (Sect. 2.5, /2.6/) that Re{$\mathbf{E}^*\cdot\mathbf{j}$} is not the power absorption density. Above the plasma frequency Re{$\mathbf{E}^*\cdot\mathbf{j}$} contains the product of the oscillating current in the plasma wave with the constant transverse field and vice versa, which oscillates with large amplitudes even when there is no absorption at all in the metal. This fact leads to the large contribution to the integral (3.70). If the proper expression $a(z,\omega) = 4\pi\gamma j^2(z,\omega)/\omega_p^2$, (2.42), is used for a sharp surface model the result of (3.70) is rather parallel to Fig. 3.21 - *7* away from ω_p with a deep minimum at ω_p. From the failure with Re{$\mathbf{E}^*\cdot\mathbf{j}$} the impression arose that a detailed consideration of the matrix elements is necessary for understanding the measured photoyield. The most elaborate calculation was published by FEIBELMAN /3.84 , 90/. He solved Maxwell's equations together with Schödinger's equation for the electrons selfconsistently. The surface is defined by a LANG - KOHN - potential /3.95/ confining the electrons. The electrons are treated in RPA. After massy numerical work he arrived at a photoyield in good agreement with experiment (Fig. 3.21-$\mathcal{3}$). In this treatment all the surface effects mentioned at the beginning of this section are included.

In the analysis of his result, Feibelman also published the fields shown in Fig. 3.20 and interpreted, that the peak near the surface in $Im\{\xi(z)\} = Im\{[E_z(z) - E_z(inside)] / [E_z(inside) - E_z(outside)]\}$ was an indication of an additional absorption mechanism: electron-hole excitations due to breaking the translational invariance. This effect would be brought about via the matrix elements between the electron states. The fields in Fig. 3.20 derived from HD optics show the same peaks with no extra surface absorption mechanism, i.e. with a constant damping γ throughout the whole metal. There is indeed large absorption near the surface, but as the HD calculations reveal, the reason is the presence of large fields and currents near the surface, not an extra absorption mechanism. This is not clear from the RPA calculation, since it has no bulk absorption mechanism at all. It would also be impossible above ω_p to keep the interpretation of $Im\{\xi(z)\}$, which according to Feibelman is proportional to $Re\{\mathbf{E}^* \cdot \mathbf{j}\}$, as absorption density.

Most photoemission calculations do not consider the selfconsistent treatment of fields and photoexcitations as most important. They evaluate the electric field by some approximation and feed this into the matrix element calculation for the photoeffect assuming that the photocurrent does not significantly influence the field.

ENDRIZ was the first to employ the fields given by HD optics in a surface photoyield calculation /3.92/. His model defined the metal surface by an image potential. By an unclear convolution with the charge density gradient he confined the contributions to the matrix elements to the surface region and got good results (see Fig. 3.21-$\mathcal{2}$) before reliable measurements were available.

Another matrix element calculation was published by BARBERAN and INGLESFIELD /3.93/. They, too, calculate the fields by the nonlocal HD optics and the electrons in a finite step barrier model. The surface of the hydrodynamical model had to be a bit inside the finite barrier. With this surface model the result in Fig. 3.21-$\mathcal{4}$ is achieved. Above ω_p the intensities are low as they should but the maximum below ω_p is too broad.

The matrix element calculations are rather involved and therefore we have tried /3.87/ how far we can get with the assumption that the matrix element or better the effect of the transition current density in the formulation of FEIBELMAN /3.84/ is constant and the frequency dependence of the photoyield is solely due to the variation of the field. On one hand the proper expression (2.42) should be used for the absorption density, on the other hand it is important to introduce the large fields near the surface below ω_p by the softness of the surface (see Sect. 3.9). When the absorption density (2.42) is calculated for a surface with a step (parameters see Sect. 3.3) and is integrated in the surface region according to (3.70) with $\ell = 10\,\text{Å}$, curve 8 in Fig. 3.21 is obtained in good agreement with experiment. We also calculated

$$\left| \int_0^{7\text{A}} E_z(z)\ dz \right|^2 \sim Y \qquad (3.71)$$

which is strictly the result when the transition current density is a constant in a region of 7 Å near the surface and zero further inside. The result in Fig. 3.21-9 is even closer to the measurements. The formulation (3.71) is independent of any approximation for damping or energy loss. Also the result in Fig. 3.21-9 is practically independent of γ in the HD and even gives the same photoyield curve with $\gamma = 0$. In this case the surface peak in the imaginary part in Fig. 3.20 disappears. Therefore the frequency dependence of the electric field explains the frequency dependence of the photoyield around ω_p.

The HD does even produce the narrower maxima for the yield from the electronic surface states /3.90/, when one confines the integral (3.70) more closely to the surface by choosing $\ell = 5$ Å /3.87/.

We have shown that the dependence on frequency of the electric field near the surface, not matrix element effects are the determining factor for the photoemission yield near ω_p of a simple metal. These fields can be simply calculated by HD optics. An approximation of the soft density gradient must, however, be included in the surface model, because the peak in $Y(\omega)$ below the plasma frequency is caused by large fields built up by the collective plasma wave, which starts already in the low density tail of the decreasing electron density at the surface. As discussed in Sect. 3.4, for the strong enhancement of the field below ω_p, it is necessary to have a standing wave resonance in the low density tail of the surface electron profile /3.32/. Recent microscopic calculations which will be discussed in Sect. 4.4 confirm these results.

3.11 Different Additional Boundary Conditions

We have discussed in Sects. 2.3 , 4 , 6 that a macroscopic approach to the surface response with surface or interface models characterized by a discontinuous change of the model parameters requires additional boundary conditions (ABCs). As we argued in Sect. 2.3 the choice of the boundary conditions should finally be judged by their success to predict in many different circumstances the right phases and amplitudes of the fields a distance away from the interface and possibly also near the interface. The best choice according to these criteria is not completely settled. The conditions (2.31 , 45 , 52) are those most widely tested in comparison to experiments as well as to microscopic calculations.

Chapter 4 deals with a number of boundary descriptions, which have been proposed as a conceptual possibility but have often not really been tested for a variety of consequences. Really competitive to those used here are probably only the ABCs of BOARDMAN and RUPPIN /2.24/. Boardman in his earlier papers /3.30 , 31/ has used in addition to (2.45c) the continuity of the normal stress at a boundary, i.e. of the hydrostatic electron gas pressure perturbation:

$$\text{continuity of } \delta p = \frac{m}{e} \beta \, \delta \rho \qquad\qquad (3.72)$$

in addition to

$$\text{continuity of } j_n = \sigma_\perp E_{\perp,n} + \sigma_\parallel E_{\parallel,n} \quad . \qquad\qquad (3.73)$$

Both requirements may sound reasonable, but at a free metal-vacuum surface only one of them can be fulfilled. How to choose? SCHWARTZ and SCHAICH /3.96/ have recently shown that with condition (3.72) there is no plasmon excitation in a metal film modelled with a sharp density step at the surface. The requirement $\delta p = 0 = \delta \rho$ at both surfaces wipes out any charge density in the metal and therefore does not show the resonances in reflection and transmission predicted by MELNYK and HARRISON /3.5/ and seen experimentally (/3.6 , 7/, Sect. 3.2). Schaich and Schwartz argue that this failure does not rule out condition (3.72) at a free surface, because the experiment was performed with layers on a substrate. If the system is made asymmetric, plasmons are excited. But in this case the resonances in reflection appear at every standing wave resonance with (3.72) /3.97/ and not at every second resonance as predicted by Melnyk and Harrison using condition (3.73). Also in the experiments it was found that the interpretation of the structures as being related to every second resonance was in agreement with the plasmon dispersion (Sect. 3.2 , 4). We consider this sequence of arguments a decision for condition (3.73) and against (3.72) at a free metal surface. This agrees with the choice taken by BOARDMAN /3.13/. With condition (3.73) a free metal layer and a layer on a substrate behave qualitatively in the same way, also in a free layer plasmon excitation is possible.

At an interface between two spatially dispersive media two additional boundary conditions are needed (Sect. 2.6). FORSTMANN and STENSCHKE /2.6/ have pointed out, that the two conditions (3.72 , 73) are not compatible with another plausible requirement: Continuity of the normal component of the energy current (2.39):

$$\text{continuity of } (4\pi\beta/\omega_p^2)\rho j_n = v_n \cdot \delta p \quad . \qquad\qquad (3.74)$$

v_n is the component of the electron mean velocity normal to the surface. (3.74) can be derived from the general statement, that all fields, charge densities and current densities should be finite, that our model should not contain singular surface densities. These requirements yield the conditions (2.45) as a mathematical consequence, especially (2.45c,d). BOARDMAN and RUPPIN /2.24/ accepted (3.74) but found a different separation into linear conditions more "natural":

continuity of v_n (3.75)

continuity of $\delta p = \frac{m}{e} \beta \, \delta \rho$. (3.76)

These are the conditions for the electrodynamics of gaseous plasmas with mobile positive and negative charges. These conditions also apply at mechanical interfaces, which are determined by the mobile materials themselves, a water-oil interface for instance. As a consequence the surfaces and interfaces move around when perturbations hit them. This movement is approximately condensed into singular surface densities, singular sinks and sources on a planar surface model, which in Boardman's treatment revive the singular surface charge densities. It could be argued, that if the model contains singular surface densities at all, also (3.74) is not necessarily true.

Our conditions (2.45c,d) which avoid singular densities alltogether, can be compared to a different mechanical analogy: The motion of an incompressible liquid in a tube which contains an interface because of a discontinuous change of the cross section. At this interface the current is continuous, but according to Bernoulli's law the velocity as well as the pressure changes discontinuously. To us this model appears as the better mechanical analogy to a metal-metal interface, because the electron liquid itself does not define the interface which is determined by immobile ion densities, "external forces" like the different tube cross sections in the analogy. At the *free surface* it appears to be the wrong approximation that the pressure perturbation of the electron liquid goes to zero, but instead the ion background takes up stresses and the current can go to zero, can be continuous. Then it is only consistent to expect the same condition at an interface.

These arguments can only serve to make our choice of boundary conditions appear at least as "natural" as those of Boardman and Ruppin. We agreed already in Sect. 2.3 that only the success is the ultimate criterion for the usefulness of an approximate surface model expressed by boundary conditions. The boundary conditions (3.72, 73) have been compared to conditions (2.45) by FORSTMANN and STENSCHKE /2.28/ in the case of the surface plasmon dispersion on a two step density surface. The prediction of the two methods were not qualitatively different in this case. A systematic study leading to differences which can be seen experimentally in one case or another or leading to inconsistencies for different predictions is still to come.

Recently SCHWARTZ and SCHAICH /3.96/ calculated a low lying phonon like eigenmode in the soft surface cavity (discussed in Sect. 3.4) by the use of Boardman's boundary conditions (3.75 , 76) at the inner interface. This mode shows up only for a double step model and not for more realistic density profiles. Like in the case of the reflection coefficient, here again the continuity of the pressure (3.75 , 76) gives results qualitatively different for "near by" models. Such an effect has never been seen with the ABCs (2.45), it is definitely not the case in the two examples studied by Schwartz and Schaich. Schwartz and Schaich were irritated by their results to a degree that they express a warning against the use of the HD /3.96/. By using ABCs (2.45) or (2.52) we never encountered this kind of problems. We always found, that neighbouring models yielded neighbouring results and reasonable approximate models yielded results in excellent comparison with those of more sophisticated treatments of the metal surface.

4. Theoretical Concepts and Models of Metal Surface Response

4.1 Additional Boundary Conditions or Susceptibility

The hydrodynamic model in conjunction with additional boundary conditions as pre - sented in Chap. 2 is one way to treat optical problems at metal surfaces. Due to its simplicity, transparency, and flexibility this method has been widely used for the interpretation of experiments, as shown in Chap. 3. In this chapter we will discuss other attempts to calculate electromagnetic fields at metal surfaces. A response theory on an intermediate level, which describes measurable response prop- erties such as reflectivity, absorptance etc. in terms of certain surface response functions, but which does not try to calculate surface electromagnetic fields, will be discussed in Chap. 5.

Attempts to calculate surface electromagnetic fields have been made by many authors with many different model assumptions, mostly on a phenomenological, but also on a microscopic level. The microscopic theory introduces the surface on the quantum mechanical level of wavefunctions. Linear response theory then yields the electromagnetic response of the system with surface in terms of a certain suscepti- bility, e.g. in the form

$$\mathbf{j}(\mathbf{r},\omega) = \int d^3 r' \; \overset{\leftrightarrow}{\sigma}(\mathbf{r},\mathbf{r}',\omega) \; \mathbf{E}(\mathbf{r}',\omega) \quad , \tag{4.1}$$

with an explicit expression for the nonlocal conductivity $\overset{\leftrightarrow}{\sigma}$. Together with Maxwell's equations, the material equation (4.1) yields a set of coupled integro-differential equations, which determine uniquely the response to incoming electromagnetic radia- tion. Microscopic caculations, to be discussed in Sect. 4.4, require a substantial amount of computational work.

The phenomenological approaches attempt to use the knowledge of the bulk re- sponse. This leads to the long standing problem of additional boundary conditions (ABC) in phenomenological optics, which has been a matter of debate since PEKAR's /2.18 , 4.1/ proposal in the context of excitonic polarization in semiconductors. According to the manner in which the bulk response properties are introduced, the need for ABC becomes apparent or not. Some authors /2.19 - 21 , 4.2/ claimed that it is not necessary to worry about ABC, because mathematics automatically solves the problem. They argued that, if the homogeneous bulk medium responds according to

$$\mathbf{j}(\mathbf{r},\omega) = \int d^3 r' \, \overset{\leftrightarrow}{\sigma}{}^h(\mathbf{r} - \mathbf{r}';\omega) \, \mathbf{E}(\mathbf{r}',\omega) \quad , \tag{4.2}$$

the response of the system containing the medium only in the halfspace $z > 0$ is given by the same expression (4.2) provided both $\mathbf{r}$ and $\mathbf{r}'$ are restricted to that halfspace. For this surface model, which was termed the "dielectric approximation", the susceptibility is obtained from the corresponding bulk susceptibility simply by truncation,

$$\overset{\leftrightarrow}{\sigma}{}^{DA}(\mathbf{r},\mathbf{r}';\omega) = \theta(z)\,\theta(z')\,\overset{\leftrightarrow}{\sigma}{}^h(\mathbf{r} - \mathbf{r}';\omega) \quad , \tag{4.3}$$

with $\theta(z) = 1$ for $z > 0$, $\theta(z) = 0$ for $z < 0$. BISHOP and MARADUDIN /2.22/ pointed out that, in the case of a spatially dispersive semiconductor, the model (4.3) leads to a discontinuity of the energy current at the surface, which then acts as a source or sink of energy. They tried to repair this defect by introducing additional surface forces.

In fact, (4.3) is not a necessary mathematical consequence of (4.2). A whole class of susceptibilities for the surface problem,

$$\overset{\leftrightarrow}{\sigma}{}^{(U)}(\mathbf{r},\mathbf{r}';\omega) = \theta(z)\,\theta(z')\,[\overset{\leftrightarrow}{\sigma}{}^h(\mathbf{r} - \mathbf{r}';\omega) + U\overset{\leftrightarrow}{\sigma}{}^h(\mathbf{r} - \overset{\leftrightarrow}{\alpha}\mathbf{r}';\omega)\overset{\leftrightarrow}{\alpha}] \quad , \tag{4.4}$$

where U is a model parameter and

$$\overset{\leftrightarrow}{\alpha} = \begin{bmatrix} 1 & 0 & 0 \\ 0 & 1 & 0 \\ 0 & 0 & -1 \end{bmatrix} \tag{4.5}$$

describes reflection at the surface, is compatible with the bulk formula (4.2) and has been considered in the literature /4.3 - 8/. As will be discussed in detail in Sect. 4.3, the special value $U = +1$ can be physically motivated by specular reflection arguments, and implies simple and reasonable boundary conditions for currents and fields. In general, the choice of a model susceptibility uniquely determines the response of the system and, consequently, the behaviour of currents and fields at the surface, although it may not be possible to express this behaviour in terms of simple boundary conditions (e.g. for $U \neq \pm 1$). If boundary conditions are imposed by physical requirements, the admissible susceptibilities will be restricted.

The relation between ABC and susceptibility becomes especially transparent, if the information about the response of the homogeneous bulk medium is not directly taken from the real space material equation (4.2), but from its Fourier transform $\mathbf{j}(\mathbf{k},\omega) = \overset{\leftrightarrow}{\sigma}{}^h(\mathbf{k},\omega)\mathbf{E}(\mathbf{k},\omega)$. The latter is formally equivalent with an equation of the type

$$\overset{\leftrightarrow}{W} \mathbf{j}(\mathbf{r},\omega) = \mathbf{E}(\mathbf{r},\omega) \tag{4.6}$$

in real space, where $\overset{\leftrightarrow}{W}$ is a differential operator containing spatial derivatives. As an example, the hydrodynamic model (2.15) can be cast into this form with

$$\overset{\leftrightarrow}{W}_{HD} \mathbf{j} = \frac{4\pi}{i\omega\omega_p^2} [\omega(\omega + i\gamma)\mathbf{j} + \beta\boldsymbol{\nabla}(\boldsymbol{\nabla} \cdot \mathbf{j})] \quad . \tag{4.7}$$

The nonlocal conductivity $\overset{\leftrightarrow}{\sigma}(\mathbf{r},\mathbf{r}';\omega)$ is an inverse of the operator $\overset{\leftrightarrow}{W}$, i.e. a Green's function of (4.6). Since in general the corresponding homogeneous equation [(4.6) for $\mathbf{E} = 0$] has non-trivial solutions, boundary conditions are necessary to uniquely determine the Green's function, i.e. the conductivity. In the homogeneous bulk case, these are the conditions of outgoing plane waves, which are automatically taken into account by the Fourier transform, and lead to (4.2).

In the form (4.6) it is clear how to use the bulk response properties right up to the surface: all the material parameters defining $\overset{\leftrightarrow}{W}$ [e.g., ω_p, γ, β in (4.7)] shall have their bulk values up to the surface plane. To complete the surface model, boundary conditions (the ABC) must be specified and then the Green's function $\overset{\leftrightarrow}{\sigma}(\mathbf{r},\mathbf{r}';\omega)$ for the surface problem is uniquely defined. As an example we consider in the following Sect. 4.2 a generalization of the hydrodynamic model which allows to satisfy alternatively two different sets of reasonable ABC leading to different results for $\overset{\leftrightarrow}{\sigma}(\mathbf{r},\mathbf{r}';\omega)$. The result for Pekar's ABC, $\mathbf{j}(z = 0^+) = 0$, is not of the form (4.4) and cannot be expressed in terms of the bulk conductivity $\overset{\leftrightarrow}{\sigma}^h(\mathbf{r} - \mathbf{r}';\omega)$.

The mentioned phenomenological approaches both have their merits and limitations. The hydrodynamic and similar approximations are easy to handle, and (4.6) together with Maxwell's equation can be solved, even without explicit evaluation of the Green's function, by a suitable ansatz of partial waves (cf. Chap. 2). The method is also easily adapted to metal-metal interfaces. However, the method works only if the differential operator $\overset{\leftrightarrow}{W}$ in (4.6) is sufficiently simple. If the bulk conductivity has a complicated analytical dependence on the wavevector $\mathbf{k}$, as , e.g., the Lindhard function which contains branch cuts, the method cannot be applied. The specular reflection model and related models, (4.4), are, on the other hand, free from this limitation and work for any bulk response function. They are, however, not easily generalized to metal-metal interfaces, although attempts in this direction have recently been made /4.9/. Moreover, there are sets of physically reasonable ABC, which cannot be satisfied by this method.

In this Chap. 4 we focus attention on the free vacuum-metal surface, and, in the phenomenological discussion, Sects. 4.2, 3, on a sharp, single-stepped surface. Hydrodynamic models of the form (4.6) with position-dependent material parameters interpolating smoothly between the metal and vacuum side have also been considered /3.25 - 27/. These models usually require a lot of computational work, already for the evaluation of the material equation, and we will not discuss them further.

4.2 Green's Functions for an Extended Hydrodynamic Model

In this section we consider a hydrodynamic model with spatial dispersion in both the longitudinal and the transversal response function. This will enable us to discuss different sets of reasonable additional boundary conditions at the free surface. The aim of this section is to clarify the relation between ABC and nonlocal conductivity of the surface problem, not a direct application to experiments. Although the model has been discussed in the context of metal optics /4.9/, there is no experimental evidence that spatial dispersion in the transversal response function is important for metals.

We write the model in the form (4.6), with

$$\overset{\leftrightarrow}{W}\mathbf{j} = \frac{4\pi}{i\omega\omega_p^2}\left\{[\omega(\omega + i\gamma) - \omega_0^2]\mathbf{j} + \beta_\ell\nabla(\nabla\cdot\mathbf{j}) - \beta_t\nabla\times(\nabla\times\mathbf{j})\right\} \quad , \tag{4.8}$$

where $\omega_0 = 0$. With $\mathbf{j} = -i\omega\mathbf{P} = \partial\mathbf{P}/\partial t$, (4.6 , 8) are the continuum approximation of the equation of motion of interacting charged oscillators in an electric field, which has been discussed by several authors /2.18 , 22 ; 4.3 , 10/. The meaning of the terms in the curly brackets is: acceleration, friction, restoring force ($\omega_0 \to 0$ for free electrons), and pressure and shear force, respectively.

For the homogeneous bulk system, Fourier transformation of (4.6 , 8) leads to a conductivity tensor with the cartesian components

$$\sigma_{\mu\nu}^h(\mathbf{k},\omega) = \sigma_t(k,\omega)(\delta_{\mu\nu} - \frac{k_\mu k_\nu}{k^2}) + \sigma_\ell(k,\omega)\frac{k_\mu k_\nu}{k^2} \quad , \tag{4.9}$$

given in terms of the longitudinal (ℓ) and transversal (t) conductivities

$$\sigma_{\ell,t}(k,\omega) = \frac{i\omega}{4\pi}\frac{\omega_p^2}{\omega(\omega + i\gamma) - \beta_{\ell,t}k^2} \quad . \tag{4.10}$$

The poles of these conductivities determine eigensolutions of the homogeneous equation $\overset{\leftrightarrow}{W}\mathbf{j} = 0$. According to the notation of Chap. 2, we specify the $\mathbf{r}$ dependence of all currents and fields to be of the form $\mathbf{j}(\mathbf{r}) = \mathbf{j}(z)\exp(ik_x x)$. With this restriction, the eigensolutions can be written as

$$\mathbf{j}^{\ell,\tau}(z) = \begin{pmatrix} k_x \\ 0 \\ \tau\lambda_\ell \end{pmatrix}e^{i\tau\lambda_\ell z} \quad , \quad \mathbf{j}^{t,\tau}(z) = \begin{pmatrix} -\lambda_t \\ 0 \\ \tau k_x \end{pmatrix}e^{i\tau\lambda_t z} \quad ,$$

$$\tag{4.11}$$

$$\mathbf{j}^{s,\tau}(z) = \begin{pmatrix} 0 \\ \lambda_t \\ 0 \end{pmatrix}e^{i\tau\lambda_t z} \quad ,$$

where $\tau = \pm 1$ refers to outgoing waves for $z \to \tau\cdot\infty$,

$$\lambda_r = [\omega(\omega + i\gamma)/\beta_r - k_x^2]^{1/2} \qquad \text{for } r = \ell, t \quad , \tag{4.12}$$

with $\mathrm{Re}\{\lambda_r\} \geq 0$ and $\mathrm{Im}\{\lambda_r\} \geq 0$, and s refers to s polarization. The modes $\mathbf{j}^{\ell,\tau}$ and $\mathbf{j}^{t,\tau}$ have p polarization, $\mathbf{j}^{\ell,\tau}$ is longitudinal, and $\mathbf{j}^{t,\tau}$ and $\mathbf{j}^{s,\tau}$ are transversal.

With $\boldsymbol{\nabla} = (ik_x, 0, \partial_z)$ the differential operator $\overset{\leftrightarrow}{W}$, (4.8), contains only z derivatives. A Green's function $\overset{\leftrightarrow}{\sigma}$ of $\overset{\leftrightarrow}{W}$ is defined by

$$\overset{\leftrightarrow}{W}(k_x, \partial_z) \, \overset{\leftrightarrow}{\sigma}(z, z', k_x; \omega) = \overset{\leftrightarrow}{1} \, \delta(z - z') \tag{4.13}$$

together with a set of boundary conditions.

For the homogeneous metal, $\overset{\leftrightarrow}{\sigma}{}^h(z, z')$ must satisfy the boundary conditions for outgoing waves, running towards $z \to +\infty$ for $z > z'$ and towards $z \to -\infty$ for $z < z'$. Since for $z \neq z'$ the columns of $\overset{\leftrightarrow}{\sigma}$ must satisfy the homogeneous differential equation (4.13) defining the eigenmodes (4.11), these columns are linear combinations of the eigenmodes satisfying the correct boundary conditions,

$$\sigma_{\mu\nu}^h(z, z', k_x; \omega) = \sum_{r=\ell, t, s} \left\{ \theta(z - z') j_\mu^{r,+}(z) a_\nu^r(z') + \theta(z' - z) j_\mu^{r,-}(z) b_\nu^r(z') \right\}. \tag{4.14}$$

The coefficients a_ν^r, and b_ν^r, are determined by the singular behaviour for $z = z'$ required by (4.13). Their calculation is straightforward. The result can be written in the form

$$\sigma_{\mu\nu}^h(z - z', k_x; \omega) = \sigma_{\mu\nu}^{(t)} + \sigma_{\mu\nu}^{(\ell)} \tag{4.15}$$

with

$$\overset{\leftrightarrow}{\sigma}{}^{(t)}(z - z', k_x; \omega) = \frac{\omega_p^2}{8\pi(\omega + i\gamma)} \begin{bmatrix} \lambda_t & 0 & -k_x \mathrm{sgn}(z - z') \\ 0 & \lambda_t + k_x^2/\lambda_t & 0 \\ -k_x \mathrm{sgn}(z - z') & 0 & k_x^2/\lambda_t \end{bmatrix} e^{i\lambda_t |z - z'|} \tag{4.16}$$

and

$$\overset{\leftrightarrow}{\sigma}{}^{(\ell)}(z - z', k_x; \omega) = \frac{\omega_p^2}{8\pi(\omega + i\gamma)} \begin{bmatrix} k_x^2/\lambda_\ell & 0 & k_x \mathrm{sgn}(z - z') \\ 0 & 0 & 0 \\ k_x \mathrm{sgn}(z - z') & 0 & \lambda_\ell \end{bmatrix} e^{i\lambda_\ell |z - z'|}. \tag{4.17}$$

For the homogeneous case there is a simpler way to the same result, namely evaluation of the one-dimensional Fourier transform

$$\sigma_{\mu\nu}^h(z - z', k_x; \omega) = \int_{-\infty}^{+\infty} \frac{dk_z}{2\pi} e^{ik_z(z - z')} \sigma_{\mu\nu}^h(\mathbf{k}, \omega) \quad , \tag{4.18}$$

using (4.9, 10) for $k_y = 0$.

If spatial dispersion of the transversal conductivity is neglected, $\beta_t \to 0$, $\lambda_t \to \infty$, as in the hydrodynamic model discussed in the preceding chapters, $\overset{\leftrightarrow}{\sigma}(t)$ becomes local. Since, for $\text{Im}\{\lambda\} > 0$,

$$-i\lambda \exp(i\lambda|z|) \to 2\delta(z) \quad \text{for} \quad |\lambda| \to \infty \quad , \tag{4.19}$$

(4.16) reduces in this limit to

$$\overset{\leftrightarrow}{\sigma}{}^{(t)}(z-z',k_x;\omega) = \frac{i\omega_p^2}{4\pi(\omega+i\gamma)}\begin{bmatrix} 1 & 0 & 0 \\ 0 & 1 & 0 \\ 0 & 0 & 0 \end{bmatrix}\delta(z-z') \quad . \tag{4.20}$$

If spatial dispersion is neglected completely, $\beta_\ell \to 0$, (4.15) reduces to the local Drude conductivity

$$\sigma_{\mu\nu}^{loc}(\omega) = i\omega_p^2\delta_{\mu\nu}\delta(z-z')/[4\pi(\omega+i\gamma)] \quad .$$

We now turn to the surface problem and replace the outgoing wave modes $\mathbf{j}^{r,-}(z)$ in (4.14) by eigenmodes which satisfy reasonable boundary conditions at the surface. To fix three new eigenmodes ($r = \ell,t,s$), we need three conditions. In view of the discussion in Chap. 2, we take $j_z(0) = 0$ as one condition. In the limit $\beta_t = 0$, when no transverse eigenmodes exist, only this condition survives. For $\beta_t \neq 0$, two additional boundary conditions can be obtained from energy considerations. FORSTMANN /4.10/ has investigated dispersive media including shear forces and shown that the energy current is continuous at the surface, if, in addition to the normal component j_z, either the parallel components j_x, j_y of the current density or their normal derivatives vanish. Then we may consider two sets of physically reasonable ABC, those corresponding to Pekar's proposal,

$$\text{P:} \quad j_\mu(z = 0^+) = 0 \quad \text{for} \quad \mu = x,y,z \quad , \tag{4.21}$$

or the "specular reflection" ABC

$$\text{SR:} \quad j_z(0^+) = 0 \quad , \quad \frac{\partial j_\mu}{\partial z}(0^+) = 0 \quad \text{for} \quad \mu = x,y \quad . \tag{4.22}$$

The conditions (4.22) are obviously satisfied by any vector field which has mirror symmetry with respect to the plane $z = 0$, $\mathbf{j}(\overset{\leftrightarrow}{\alpha}\mathbf{r}) = \overset{\leftrightarrow}{\alpha}\mathbf{j}(\mathbf{r})$ with $\overset{\leftrightarrow}{\alpha}$ given by (4.5), and continuous first derivatives.

The ABC (4.22) are satisfied by the even linear combinations of the eigenmodes (4.11),

$$\mathbf{j}^{r,e}(z) = \mathbf{j}^{r,+}(z) + \mathbf{j}^{r,-}(z) \quad , \quad r = \ell,t,s \quad . \tag{4.23}$$

Inserting $\mathbf{j}^{r,e}(z)$ instead of $\mathbf{j}^{r,-}(z)$ into (4.14), we can evaluate the conductivity for specular reflection ABC. The result can be written in terms of the bulk conductivity (4.15) as (for $z,z' > 0$)

$$\overleftrightarrow{\sigma}^{SR}(z,z',k_x;\omega) = \overleftrightarrow{\sigma}^{h}(z - z',k_x;\omega) + \overleftrightarrow{\sigma}^{h}(z + z',k_x;\omega)\overleftrightarrow{\alpha} \quad . \tag{4.24}$$

This is of the form (4.4) with $U = 1$, except that the x, y dependence is Fourier transformed, with $k_y = 0$. The structure (4.24) of the susceptibility thus is, within the model (4.8), a consequence of the ABC (4.22).

JOHNSON and RIMBEY /4.3/ have investigated the question, whether the boundary conditions (4.22) are in turn a mathematical consequence of the structure (4.24) of the susceptibility. The answer is yes, provided the homogeneous-bulk susceptibility is sufficiently well behaved. Using the explicit results (4.15 - 17), one derives from (4.24) indeed easily the boundary condition (4.22). If, however, the "metal model" of Chap. 2 is considered, which is the present model with $\beta_t = 0$, the transverse contribution (4.16) is replaced by the singular term (4.20), which contributes only to the first term on the right hand side of (4.24), not to the second. As a consequence, for the metal model the structure (4.24) yields only $j_z(0^+) = 0$, whereas $\partial j_x/\partial z$ and $\partial j_y/\partial z$ will not vanish at the surface.

This example demonstrates that the question of boundary conditions implied by the structure (4.24) of the susceptibility is closely related to the analytical behaviour of the bulk susceptibility $\overleftrightarrow{\sigma}^{h}(z - z')$ for small values of $z - z'$ or, equivalently, to the large-k behaviour of its Fourier transform. We come back to this question in the following Sect. 4.3.

The even modes (4.23) and the odd modes

$$\mathbf{j}^{r,0}(z) = \mathbf{j}^{r,+}(z) - \mathbf{j}^{r,-}(z) \quad , \qquad r = \ell,t,s \quad , \tag{4.25}$$

can be used to construct three linear combinations of eigenmodes (4.11) which satisfy Pekar's ABC (4.21):

$$\mathbf{j}^{s,0} \quad , \quad \mathbf{j}^{(1)} = \lambda_\ell \mathbf{j}^{t,e} - k_x \mathbf{j}^{\ell,0} \quad , \quad \mathbf{j}^{(2)} = k_x \mathbf{j}^{t,0} + \lambda_t \mathbf{j}^{\ell,e} \quad . \tag{4.26}$$

Note that for p polarization transverse and longitudinal modes must be mixed in order to satisfy (4.21). Inserting the modes (4.26) instead of $\mathbf{j}^{r,-}$ into (4.14), one calculates a conductivity tensor which can be written in the form

$$\overleftrightarrow{\sigma}^{P}(z,z',k_x;\omega) = \overleftrightarrow{\sigma}^{h}(z - z',k_x;\omega) - \overleftrightarrow{\alpha}\,\overleftrightarrow{\sigma}^{h}(z + z',k_x,\omega)\overleftrightarrow{\alpha} + \overleftrightarrow{Q}(z,z',k_x;\omega) \quad . \tag{4.27}$$

The last term,

$$\overleftrightarrow{Q}(z,z',k_x;\omega) =$$

$$\frac{\omega_p^2 k_x}{4\pi(\omega + i\gamma)} \cdot \frac{e^{i\lambda_t z} - e^{i\lambda_\ell z}}{k_x^2 + \lambda_t \lambda_\ell}
\begin{bmatrix}
k_x \lambda_t (e^{i\lambda_t z'} - e^{i\lambda_\ell z'}) & 0 & k_x^2 e^{i\lambda_t z'} + \lambda_t \lambda_\ell e^{i\lambda_\ell z'} \\
0 & 0 & 0 \\
\lambda_t \lambda_\ell e^{i\lambda_t z'} + k_x^2 e^{i\lambda_\ell z'} & 0 & k_x \lambda_\ell (e^{i\lambda_t z'} - e^{i\lambda_\ell z'})
\end{bmatrix} \tag{4.28}$$

cannot be expressed in terms of the homogeneous-bulk susceptibility, and cannot be written as a sum of terms which depend only on either $z - z'$ or $z + z'$.

JOHNSON and RIMBEY /4.3/ asserted that the susceptibility for Pekar's ABC is determined by the first two terms on the right hand side of (4.27), without the term $\overset{\leftrightarrow}{Q}$. In general, this is not correct, since the truncated expression (without $\overset{\leftrightarrow}{Q}$) does not satisfy (4.13), i.e., is not compatible with the bulk material equation. The correction term $\overset{\leftrightarrow}{Q}$ vanishes for two special cases of little interest: first, for normal incidence, $k_x = 0$, and arbitrary values of β_t and β_ℓ, second, for $\beta_t = \beta_\ell$, which implies $\lambda_t = \lambda_\ell$, and arbitrary k_x. In both cases the surface susceptibility tensor is diagonal. Only in these special cases the discussion of Johnson and Rimbey concerning Pekar's ABC is applicable.

Again it is interesting to consider the limit $\beta_t \to 0$, i.e. the "metal model". Whereas (4.15 - 17) and (4.27, 28) immediately yield $\mathbf{j}(0^+) = 0$ for finite values of β_t and β_ℓ, in the limit $\beta_t \to 0$, $\lambda_t \to \infty$, $\exp(i\lambda_t z) \to 0$, $\overset{\leftrightarrow}{\sigma}^P$ and $\overset{\leftrightarrow}{\sigma}^{SR}$ become identical and only the boundary condition $j_z(0^+) = 0$ survives. If spatial dispersion is neglected in the transverse response, no further ABC for the parallel components j_x, j_y can be satisfied.

Whereas, within the model (4.8), the conductivity formulas (4.24 , 27) for $\overset{\leftrightarrow}{\sigma}^{SR}$ and $\overset{\leftrightarrow}{\sigma}^P$ are equivalent with simple sets of ABC for the current density, the "dielectric approximation" (4.3) implies boundary conditions which are not so easily formulated and depend on the involved eigenmodes (4.11), as was noticed by JOHNSON and RIMBEY /4.3/. In special cases, e.g. when $|k_x| \ll |\lambda_\ell|$, $|\lambda_t|$, or when $\lambda_t \approx \lambda_\ell$, the conductivity tensor $\overset{\leftrightarrow}{\sigma}^{DA}$ is diagonal and implies for instance

$$\frac{\partial j_z}{\partial z}(0^+) + i\lambda_\ell j_z(0^+) = 0 \quad , \tag{4.29}$$

$$\frac{\partial j_\mu}{\partial z}(0^+) + i\lambda_t j_\mu(0^+) = 0 \quad , \quad \text{for } \mu = x,y \quad . \tag{4.30}$$

In general, the eigenmodes mix, $\overset{\leftrightarrow}{\sigma}^{DA}$ is non-diagonal, and the implied ABC cannot be written in the form $A\partial j_\mu/\partial z + Bj_\mu = 0$. For the "metal model" only (4.29) survives. RAJAGOPAL and FORSTMANN /4.11/ discussed this situation and emphasized that in the "dielectric approximation" the surface acts as a source or sink for the electrical current and the energy current. With metals, (4.29) has not been found a useful boundary condition.

In summary, the same material equation of the form (4.6), e.g. (4.8), leads to different surface response functions $\overset{\leftrightarrow}{\sigma}(z,z',k_x;\omega)$ if different sets of ABC are employed. There are sets of ABC, e.g., the specular reflection ABC (4.22) which allow to express the surface response function in terms of the bulk response function $\overset{\leftrightarrow}{\sigma}^h(z - z',k_x;\omega)$, and other, e.g. Pekar's ABC (4.21), for which this is not possible.

On the other hand, a given expression $\overleftrightarrow{\sigma}(z,z';k_x,\omega)$ determines the behaviour of the current density and fields at the surface completely. It is, however, not always possible to specify this behaviour in terms of a simple set of ABC.

4.3 The Specular Reflection Model

A widely used phenomenological surface model on a microscopic level states that the conduction electrons are specularly reflected at the surface plane /3.1 ; 4.4 , 12/. This system responds like one side of a homogeneous system with a mirror plane for all properties and also for the exciting fields. On a macroscopic level this means that the current density $\mathbf{j}$ and the electric field $\mathbf{E}$ in the metal halfspace of interest ($z > 0$) can be considered as the current density $\mathbf{j}^{eff}$ and electric field $\mathbf{E}^{eff}$ in an effective homogeneous system submitted to symmetry conditions of the form

$$\mathbf{E}^{eff}(z) = \begin{cases} \mathbf{E}(z) & , \quad z > 0 \\ \overleftrightarrow{\alpha}\mathbf{E}(-z) & , \quad z < 0 \end{cases} , \tag{4.31}$$

where $\overleftrightarrow{\alpha}$ is the specular reflection matrix defined in (4.5). With these constraints on the solutions of Maxwell's equations, the effective homogeneous system responds with the bulk conductivity,

$$\mathbf{j}^{eff}(\mathbf{r}) = \int_{all\ space} d^3r'\ \overleftrightarrow{\sigma}^h(\mathbf{r} - \mathbf{r}';\omega)\ \mathbf{E}^{eff}(\mathbf{r}') \quad . \tag{4.32}$$

Due to (4.31), this can for $z > 0$ be written in the form

$$\mathbf{j}(\mathbf{r}) = \int_{z'>0} d^3r'\ \overleftrightarrow{\sigma}^{SR}(\mathbf{r},\mathbf{r}';\omega)\ \mathbf{E}(\mathbf{r}') \quad , \tag{4.33}$$

where the surface response function is given in terms of the bulk conductivity as

$$\overleftrightarrow{\sigma}^{SR}(\mathbf{r},\mathbf{r}';\omega) = \overleftrightarrow{\sigma}^h(\mathbf{r} - \mathbf{r}';\omega) + \overleftrightarrow{\sigma}^h(\mathbf{r} - \overleftrightarrow{\alpha}\mathbf{r}';\omega)\overleftrightarrow{\alpha} \quad , \tag{4.34}$$

i.e. (4.4) for U = +1.

The result (4.34) does not depend on the actual form of the bulk conductivity $\overleftrightarrow{\sigma}^h$, and even the assumption that $\overleftrightarrow{\sigma}^h$ depends only on the difference $\mathbf{r} - \mathbf{r}'$ is not essential and can be relaxed in order to include lattice effects /4.3/. KLIEWER and FUCHS /3.83 ; 4.4 , 5 , 13/ have extensively discussed this specular reflection model and its application to various surface problems. The "SCIB" (semi-classical infinite barrier) model which has been investigated by several authors /4.14 - 16/, is obtained if the bulk response is described by Lindhard's RPA dielectric function for the electron gas.

The solution of Maxwell's equations in the effective homogeneous system with the material equation (4.32) can be obtained by Fourier transformation. Owing to (4.31), non-trivial outgoing-wave solutions $(z \rightarrow +\infty)$ of this homogeneous problem must have discontinuities at $z = 0$, the size of which determines the amplitude of the fields inside the metal /4.3 , 4/. For p polarization, the discontinuities can be expressed in terms of $D_z(z = 0^+) = -D_z^{eff}(0^-)$ which, together with the amplitude of the reflected field, must be determined from the standard matching conditions (E_x and D_z continuous) at the vacuum/metal interface. Since for the actual physical system D_z is continuous, we write $D_z(0^+) = D_z(0)$. Then the field inside the metal can be written as /4.3 , 4 , 13 , 16/:

$$E_x(z) = -\frac{i}{\pi} D_z(0) \int\limits_{-\infty}^{+\infty} \frac{dk_z}{k^2} e^{izk_z} \left\{ \frac{k_z^2}{k_x(\varepsilon_t - c^2 k^2/\omega^2)} + \frac{k_x}{\varepsilon_\ell} \right\} \quad , \tag{4.35}$$

$$E_z(z) = \frac{i}{\pi} D_z(0) \int\limits_{-\infty}^{+\infty} \frac{dk_z}{k^2} e^{izk_z} \left\{ \frac{k_z}{\varepsilon_t - c^2 k^2/\omega^2} - \frac{k_z}{\varepsilon_\ell} \right\} \quad , \tag{4.36}$$

where $k^2 = k_x^2 + k_y^2$ and the dielectric tensor of the homogeneous bulk metal has been assumed to have the form (4.9), with $\varepsilon_t = \varepsilon_t(k,\omega)$ and $\varepsilon_\ell = \varepsilon_\ell(k,\omega)$ the transverse and longitudinal bulk dielectric functions, respectively. For $k_x \neq 0$ the integrals are well defined, since the frequency arguments of the dielectric functions have a positive (infinitesimal) imaginary part, and no singularities occur on the real k_z axis /2.3 ; 4.16/. In contrast to the electric field, the displacement field within the metal $(z > 0)$, given by

$$\begin{Bmatrix} D_x(z) \\ D_z(z) \end{Bmatrix} = -\frac{i}{\pi} D_z(0) \int\limits_{-\infty}^{+\infty} dk_z \, e^{izk_z} \frac{k_z}{k^2 - \omega^2 \varepsilon_t/c^2} \begin{Bmatrix} -k_z/k_x \\ 1 \end{Bmatrix} \quad , \tag{4.37}$$

is purely transversal, $\boldsymbol{\nabla} \cdot \mathbf{D} = 0$ for $z > 0$.

JOHNSON and RIMBEY /4.3/ have discussed the boundary values of the fields and the polarization or, equivalently, the current density within the specular reflection model in some detail. They emphasized the relation between these boundary values $(z \rightarrow 0)$ and the analytical properties of the bulk susceptibility for small values of $z - z'$ or, equivalently, large k. The most important results are easily obtained /4.16/ from (4.35 - 37), using the fact that ε_t and ε_ℓ are even functions of k_z, and that

$$\lim_{z \rightarrow 0} \int\limits_{0}^{\infty} dk_z \frac{1}{k_z} \sin(zk_z) F(k_z) = \frac{\pi}{2} F(\infty) \quad . \tag{4.38}$$

Equation (4.36) yields

$$E_z(0^+) = D_z(0)/\varepsilon_\ell(k \to \infty, \omega) \quad , \tag{4.39}$$

provided $\varepsilon_t(k,\omega)/k^2$ becomes small for large values of k. If $\varepsilon_\ell(\infty,\omega) = 1$, which means that the metal electrons cannot follow a longitudinal perturbation of arbitrarily short wave length, E_z together with D_z is continuous at the surface. The metal model of Chap. 2, the extended hydrodynamic model of Sect. 4.2 [cf. (4.10)] and also the Lindhard RPA dielectric functions (see Sect. 4.5) have this property $\varepsilon_\ell(\infty,\omega) = 1$, whereas the local Drude dielectric function is independent of k and leads to a discontinuous $E_z(z)$.

For the current density $\mathbf{j} = \omega(\mathbf{D} - \mathbf{E})/4\pi i$ the boundary condition (4.39) reads

$$j_z(0^+) = \frac{\omega}{4\pi i}\left[1 - \frac{1}{\varepsilon_\ell(\infty,\omega)}\right]D_z(0) \quad . \tag{4.40}$$

Similarly one derives /4.16/

$$\frac{\partial j_x}{\partial z}(0^+) = \frac{\omega}{4\pi k_x}\left\{\frac{\omega^2}{c^2}\left[1 - \varepsilon_t(\infty,\omega)\right] + k_x^2\left[1 - \frac{1}{\varepsilon_\ell(\infty,\omega)}\right]\right\}D_z(0) \quad . \tag{4.41}$$

For the extended hydrodynamic model of Sect. 4.2 with finite shear forces, $\beta_t \neq 0$ in (4.8, 10), $\varepsilon_t(\infty,\omega) = \varepsilon_\ell(\infty,\omega) = 1$, and (4.40, 41) reduce to the specular reflection ABC (4.22). For the metal model of Chap. 2, on the other hand, the transverse dielectric function is independent of k ($\beta_t = 0$), and $\partial j_x/\partial z$ is not continuous at the surface. The same result is obtained for the RPA Lindhard function, which satisfies $\varepsilon_t(\infty,\omega) = \varepsilon_t(0,\omega) \neq 1$ (cf. Sect. 4.5).

With the dielectric functions (2.19) of the hydrodynamic model (HD) the integrals in (4.35 - 37) are easily evaluated and the specular reflection model then reduces to the HD. The longitudinal electric field originates according to (4.35, 36) from the zero of $\varepsilon_\ell(k,\omega)$, i.e. the plasmon excitation, which causes the nonlocal response within the HD. If, on the other hand, Lindhard's RPA dielectric function is inserted (SCIB model), the longitudinal field contains in addition to the plasmon term contributions from branch cuts, which describe the optical excitation of electron-hole pairs at the surface. A detailed discussion of the analytical properties of the Lindhard dielectric functions in the complex k plane and a numerical evaluation of the plasmon pole and cut contributions has been given by GERHARDTS /4.16/.

For the sake of completeness, we also give the results for s polarization. The standard matching conditions require E_y and $B_x = (ic/\omega)\partial E_y/\partial z$ continuous. The electric field inside the metal (z > 0) can be written as

$$E_y(z) = -\frac{i}{\pi}B_x(0)\int\limits_{-\infty}^{+\infty} dk_z\, e^{izk_z}\frac{c/\omega}{\varepsilon_t - c^2k^2/\omega^2} \quad , \tag{4.42}$$

and the displacement field results if an additional factor $\varepsilon_t(k,\omega)$ is inserted into the integral. The boundary condition similar to (4.41) is

$$\frac{\partial j_y}{\partial z}(0^+) = \frac{\omega^2}{4\pi c} [1 - \varepsilon_t(\infty,\omega)] B_x(0) \quad . \tag{4.43}$$

[Note that $D_z(0) = ck_x B_y(0)/\omega$ in (4.41).] The fields in the case of s polarization are purely transversal.

Phenomenological generalizations of the specular reflection model with suscepti-bilities $\overset{\leftrightarrow}{\sigma}(U)$ of the form (4.4) have also been discussed in the literature. RIMBEY and MAHAN /4.17/ proposed U = -1 ("anti-specular-reflection") to describe reflec-tion of excitons /4.18/ at semiconductor surfaces. This model simulates Pekar's ABC ($j(0^+)$ = 0) for normal incidence of light, but not otherwise (see Sect. 4.2). As has been emphasized by JOHNSON and RIMBEY /4.3/, the Rimbey - Mahan model (U = -1) excludes the possibility of optical excitation of longitudinal modes in the medium and reduces to the local Fresnel theory, if the dispersion in the transversal dielec-tric function $\varepsilon_t(k,\omega)$ is neglected. Therefore, U = -1 cannot be used for an adequate description of a metal surface.

Also models with U $\neq$ ±1 have been discussed /4.7, 8/. The solution of Maxwell's equations in the effective homogeneous system under the restriction of the "U-sym-metry" $E_x(-z) = UE_x(z)$, $E_z(-z) = -UE_z(z)$ (for z > 0) with U $\neq$ ±1 is only possible, if in the unphysical halfspace (z < 0) of the effective homogeneous system suitable source terms are introduced /4.8/. The resulting formalism is neither mathematically simple or general, nor physically cogent, since there are completely reasonable sets of ABC, e.g. Pekar's ($j(0^+)$ = 0), which cannot be simulated by·such a model. In the same spirit, the interface between two metallic halfspaces has been treated /4.9/: Symmetry conditions for effective fields are formulated which allow to map the inter-face system on effective homogeneous systems (with surface and volume charges and currents) which respond with bulk susceptibilities. The symmetry conditions are, on the other hand, interpreted in terms of the probabilities that an electron at the interface will be transmitted, specularly reflected or diffusely reflected. We don't see in which respect this type of phenomenological interface model can be helpful for an understanding of the physical effects at the interface.

4.4 Microscopic Response Theory

On a microscopic, quantum mechanical level the surface is defined by a potential which keeps the electrons inside the metal. When the electron motion is defined by a Hamiltonian describing the system with a surface, the response of the electron system to a weak external perturbation, e.g. an incident electromagnetic field, can be calculated by standard linear response theory. There is no need and no room for additional boundary conditions, which constitute a central and controversial topic of phenomenological theories.

For bulk systems, the microscopic theory of dielectric properties has been developed in great detail. The dielectric response of an interacting electron gas with a uniform background of positive charge (jellium) has been studied including exchange and correlation effects /2.3/, and also the so called local field effects due to the discrete nature of periodic solids have been investigated /2.5 ; 4.19/. At a surface, these and additional effects, e.g. surface roughness, play a role and should in principle be taken into account in calculations of surface electromagnetic fields. But up to date microscopic calculations of optical surface response are usually restricted to the simplified model of two-dimensionally translation-invariant jellium in the random phase approximation (RPA).

As is well known /2.3 , 12/, the RPA can be formulated as mean-field approximation: The mutual Coulomb interaction between electrons changes the bare incident electromagnetic field into a self-consistent field to which the electrons respond as non-interacting particles, i.e. with a conductivity tensor /2.3 ; 4.20/

$$\sigma_{\mu\nu}(\mathbf{r},\mathbf{r}';\omega) = \frac{ie^2}{m\omega} \delta_{\mu\nu} \delta(\mathbf{r}-\mathbf{r}') \, n(\mathbf{r})$$

$$- \frac{2e^2}{i\omega} \sum_{\mathbf{p},\mathbf{p}'} \frac{f(E_{\mathbf{p}}) - f(E_{\mathbf{p}'})}{E_{\mathbf{p}} - E_{\mathbf{p}'} + \hbar\omega + i0^+} \, j^{\mu}_{\mathbf{p}\mathbf{p}'}(\mathbf{r}) \, j^{\nu}_{\mathbf{p}'\mathbf{p}}(\mathbf{r}') \quad , \tag{4.44}$$

where

$$n(\mathbf{r}) = 2 \sum_{\mathbf{p}} |\phi_{\mathbf{p}}(\mathbf{r})|^2 \, f(E_{\mathbf{p}}) \tag{4.45}$$

is the electron density in the ground state, $f(E_{\mathbf{p}}) = \theta(E_F - E_{\mathbf{p}})$ the zero temperature Fermi function, and E_F the Fermi energy. The single-particle energy eigenvalues and eigenfunctions are $E_{\mathbf{p}} = \hbar^2 \mathbf{p}^2/2m$ and

$$\phi_{\mathbf{p}}(\mathbf{r}) = \exp[i(xp_x + yp_y)] \, \varphi_{p_z}(z) \quad , \tag{4.46}$$

respectively, and the current density matrix elements are

$$j^{\mu}_{\mathbf{p}\mathbf{p}'}(\mathbf{r}) = \frac{\hbar}{2mi} \left[\phi^*_{\mathbf{p}}(\mathbf{r}) \frac{\partial \phi_{\mathbf{p}'}}{\partial r_{\mu}} - \frac{\partial \phi^*_{\mathbf{p}}}{\partial r_{\mu}} \phi_{\mathbf{p}'}(\mathbf{r}) \right] \quad . \tag{4.47}$$

For optical problems it is important that the self-consistent field $\mathbf{E}$ is calculated via the full Maxwell's equations, with the material equation given by (4.1 , 44), and not in the static approximation, which does not allow for propagating fields outside the metal.

FEIBELMAN /3.84 , 85/ was the first author who really attacked this problem and presented numerical calculations of electromagnetic surface fields /3.85 , 86/ and reflectivity /3.86 ; 4.21/. He also calculated more general surface response functions /3.84 , 86 ; 4.22/ which will be discussed in more detail in Chap. 5. Here we give only a rough sketch of Feibelman's calculation procedure and refer to his original work /3.85 ; 4.21/ and to his review /3.84/ for further details of his involved analysis.

Using a standard Green's function technique and Fourier transformation parallel to the surface, FEIBELMAN /3.85/ converts the integro-differential equations formed by Maxwell's equations together with the material equation into an integral equation for the z-dependent vector potential (i.e. the electric field), with the externally imposed field as the driving inhomogeneous term. Further simplification is achieved by explicit introduction of the transverse field and by exploiting the "long-wavelength limit", which assumes that wavelength and penetration depth of the transverse light wave (being typically of the order of 10^3 Å) are much larger than the width of the surface region, in which the conductivity is intermediate between the bulk value and the zero vacuum value, and also than the decay length or wavelength of plasma waves. Then only for the normal (z) component of the electric field (which includes the longitudinal field) a nontrivial integral equation with the kernel $\sigma_{zz}(z,z'$, $k_\parallel = 0 ; \omega)$ remains to be solved numerically, whereas the evaluation of the other components, related to the transverse field only, becomes trivial. Feibelman calculates the single-particle quantities $\varphi_{p_z}(z)$ etc. determining σ_{zz} from the self- consistent effective single-particle potential calculated by LANG and KOHN /4.23/ for a jellium surface in the local density approximation of the HOHENBERG - KOHN - SHAM theory /4.24/. Having computed σ_{zz}, he solves the integral equation numerically, taking explicitly into account that far inside the metal the electric field consists of several contributions, the transverse field, slowly-decaying Friedel-type oscillations, and eventually (for $\omega > \omega_p$) a propagating plasma wave.

Feibelman's approach incorporates a realistic description of the electron density profile at the surface in the ground state and a reasonable treatment of optical response, which is given by electron-hole excitations in the RPA. His theory should be adequate for free electron metals (extension to d-electron metals such as Ag, Au, Cu has not been attempted) and indeed reproduces, for instance, the frequency dependence of the photoyield spectrum of aluminum /3.84 , 90; Sect. 3.10/. The method is, on the other hand, not very transparent and flexible and it is hard to extract physical insight from its results.

An alternative RPA treatment of electrodynamics at a metal surface has been pursued by MANIV and METIU in a series of papers /4.25 - 28/. Instead of working with Maxwell's equations, they consider directly the response of the interacting electrons to an externally applied "bare" electromagnetic field, which is given by a scalar and a vector potential. The response of the interacting system is determined

by a (4 × 4) polarization tensor, which in the RPA is related to the bare polarization tensor of the non-interacting system by a linear integral equation involving
the photon propagator (Green's function of Maxwell's wave equation) /2.3 ; 4.25/.
The components of the bare polarization tensor are density-density, density-current,
and current-current correlation functions, the latter being essentially the conductivity tensor given in (4.44). In the homogeneous bulk case the integral equation
becomes a simple algebraic equation in Fourier space. For the jellium halfspace,
Maniv and Metiu use a mixed Fourier representation, with the usual propagating waves
parallel and sinus or cosinus functions (depending on the four-vector components)
normal to the surface. To keep the expression for the polarization tensor simple,
only the "infinite barrier model" (IBM) is considered, which restricts the electrons
to the metal region by an infinite potential step at the surface /4.29/, so that the
wavefunctions normalized to a volume $0 < x,y < L$, $0 < z < \infty$ are given by (4.46) with

$$\varphi_{p_z}(z) = \frac{1}{L} \left(\frac{2}{\pi}\right)^{1/2} \sin(z p_z) \quad . \tag{4.48}$$

The approach of Maniv and Metiu has the advantage that, once the polarization tensor of the interacting system has been calculated, the response to any external
perturbation is easily obtained with no more effort than evaluating integrals. Maxwell's equations, on the other hand, must be solved anew if the incident plane wave
is replaced, for instance, with the field produced by an oscillating dipole near
the metal surface.

Practical calculations are, however, complicated by the fact that slowly-varying
long-range field contributions and rapidly-varying short-range contributions must
be treated simultaneously. The difficulties were surmounted by suitable renormaliza-
tion procedures, but the resulting formalism appears rather heavy. Nevertheless,
explicit numerical results have been obtained, e.g., for surface fields /4.25/, reflectivity and photoyield of thin films /4.26/, polarization by an oscillating dipole /4.27/, and for Raman scattering by localized vibration modes /4.28/.

A simple and transparent approach to microscopic metal optics has been proposed
by GERHARDTS and KEMPA /4.30/. Assuming that the metal electrons are confined to
the halfspace $z > 0$, the components of the electric field $\mathbf{E}(\mathbf{r},t) = \mathbf{E}(z)\exp i(xk_x - \omega t)$
and of the displacement field $\mathbf{D}$ inside the metallic halfspace are expanded according
to the mixed Fourier-transformation

$$F_\mu(z) = \frac{2}{\pi} \int_0^\infty dk_z \, \mathscr{F}_\mu(k_z)\cos(z k_z) \quad , \quad \mu = x,y \quad , \tag{4.49a}$$

$$F_z(z) = \frac{2i}{\pi} \int_0^\infty dk_z \, \mathscr{F}_z(k_z)\sin(z k_z) \quad . \tag{4.49b}$$

Taking cosine transforms of the x,y components of Maxwell's equation (2.5) and the sine transform of the z component, one obtains

$$k_z^2 \mathscr{E}_x(k_z) - k_x k_z \mathscr{E}_z(k_z) - \frac{\omega^2}{c^2} \mathscr{D}_x(k_z) = \frac{i\omega^2}{k_x c^2} D_z(0) \qquad (4.50a)$$

$$-k_x k_z \mathscr{E}_x(k_z) + k_x^2 \mathscr{E}_z(k_z) - \frac{\omega^2}{c^2} \mathscr{D}_z(k_z) = 0 \qquad (4.50b)$$

for p polarization and

$$(k_x^2 + k_z^2) \mathscr{E}_y(k_z) - \frac{\omega^2}{c^2} \mathscr{D}_y(k_z) = -\frac{\partial E_y}{\partial z}(0) \qquad (4.51)$$

for s polarization. The boundary terms on the right hand sides of these equations arise from integration by parts. These terms together with the reflection amplitudes must be determined by the standard matching conditions between the vacuum fields in the halfspace $z < 0$ and the metal fields at $z = 0$. Mixed Fourier transformation of the material equation (4.1) leads to the equivalent form

$$\mathscr{D}_\mu(k_z) = \sum_\nu \int_0^\infty dk_z' \; \varepsilon_{\mu\nu}(k_x; k_z, k_z'; \omega) \; \mathscr{E}_\nu(k_z') \quad . \qquad (4.52)$$

For the IBM with wavefunctions (4.48), the RPA formula (4.44) yields a result of the form /4.30/

$$\varepsilon_{\mu\nu}(k_x; k_z, k_z'; \omega) = \delta(k_z - k_z') \; \varepsilon_{\mu\nu}(\mathbf{k}; \omega) + N_{\mu\nu}(k_x; k_z, k_z'; \omega) \quad , \qquad (4.53)$$

where $\varepsilon_{\mu\nu}(\mathbf{k}; \omega)$ is the dielectric tensor of the homogeneous bulk jellium, which has the structure (4.9) and can easily be expressed in terms of Lindhard's RPA dielectric functions $\varepsilon_t(k, \omega)$ and $\varepsilon_\ell(k, \omega)$ (cf. Sect. 4.5 below), and where $N_{\mu\nu}$ contains no $\delta(k_z - k_z')$ singularity.

It can be shown /4.31/ that only the kernels $N_{\mu\nu}$ in (4.53) are modified if, in addition to the infinite barrier potential, a surface potential $v(z)$ is introduced (for $z > 0$) which produces a smoother, more realistic electron density profile than the IBM. The kernels $N_{\mu\nu}$ describe surface properties, notably the effect of the smooth electron density profile. The singular diagonal term, on the other hand, describes the bulk properties of the metallic halfspace and is not affected by surface properties.

If the surfaces kernels $N_{\mu\nu}$ are omitted in (4.53), the microscopic RPA model reduces to the phenomenological SCIB model, i.e. the specular reflection model with the bulk response given by Lindhard's RPA dielectric functions. Indeed, then (4.50 - 52) are easily solved, and with (4.49) the results (4.35 - 43) are recovered. So it is surface effects, not quantum effects, what is neglected in the "semi-classical infinite barrier" model. There is, however, no justification for the omission of the $N_{\mu\nu}$, if one is interested in surface effects on, for instance, the surface electromagnetic fields.

From the explicit expressions for the $N_{\mu\nu}$ within the IBM one can show /4.30/ that, in the long-wavelength limit (important $|k_z| \gg |k_x| \sim \omega/c$), longitudinal field contributions are decoupled from transverse contributions [both defined in the mixed Fourier representation with respect to the direction of $\mathbf{k} = (k_x, 0, k_z)$]. Furthermore, the transverse component of the kernel $N_{\mu\nu}$ and the spatial dispersion in $\varepsilon_t(k,\omega)$ can be neglected, so that the transverse fields are easily evaluated, and only a single one-dimensional integral equation for the longitudinal component of the electric field remains, just as in Feibelman's real-space approach. Solving the integral equation numerically, GERHARDTS and KEMPA /4.30/ calculated surface electromagnetic fields, which were found to agree with results of MANIV and METIU /4.25/, and, in addition, the frequency-dependent surface absorptance, which were found to be typically about two orders of magnitude smaller (for $\omega < \omega_p$) than corresponding results obtained by FEIBELMAN /3.84/. This discrepancy indicates that details of the electron distribution in the surface region are extremely important for the optical surface properties.

A systematic investigation of the dependence of optical response properties on the steepness of the surface profile has recently been performed by GIES and GERHARDTS /4.32/. In order to change the surface profile in a physically meaningful way, charged surfaces were considered, since with increasing positive surface charge the electron density profile becomes steeper. The surface potential was taken from a self-consistent Lang - Kohn - type calculation /4.33/ for charged surfaces, using a slab geometry with infinite potential barriers located so far away from the jellium slab, that they do not disturb the electron distribution in the ground state. This realistic generalization of the IBM can also be treated in the mixed Fourier-representation /4.34/. Figure 4.1 shows results obtained by GIES and GERHARDTS /4.32/ for the complex surface response function $d_\perp(\omega)$ which was introduced by FEIBELMAN /3.84 , 86/ and is discussed in details in Chap. 5. Its real part measures the dipole length of the optically induced charge density with respect to the jellium edge and its imaginary part is closely related to the optical absorption in the surface region. For computational reasons relatively thin slabs were considered, so that finite size oscillations are seen in curves a and b, and a phenomenological damping parameter γ was introduced (see Sect. 4.5 below). The dramatic change of surface response properties with changing surface charge revealed by Fig. 4.1 indicates excitation of a collective surface mode in sufficiently diffuse surfaces /3.32 - 34/. This effect is discussed in some detail in Sect. 3.4 and in Sect. 5.7.

In the present section we considered only microscopic treatments of metal optics which lead to explicit calculations of surface fields or surface response functions.

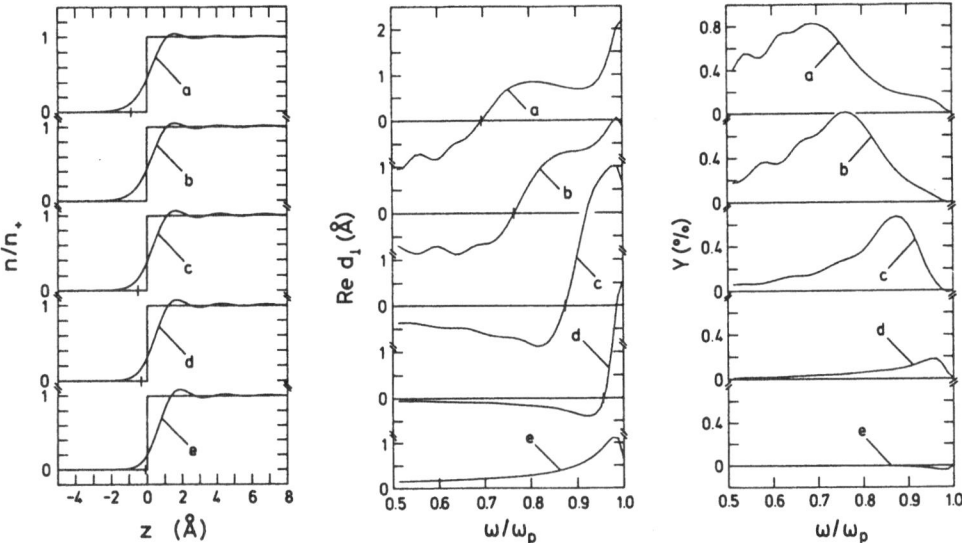

Fig. 4.1. Self-consistent ground-state electron-density n(z), and surface response functions Re{$d_\perp(\omega)$} and $Y(\omega) = \sqrt{8}(\omega/c)[(\omega/\omega_p)^2 - 1]$ Im{$d_\perp(\omega)$} for a charged jellium slab with surface-charge density σ(in 10^{-3} e/Å^2) = -2.39 (a), 0.0 (b), 4.77 (c), 11.93 (d), 23.87 (e). The electrons are confined to the interval $-5 < z/\text{Å} < 59$, the positive background with $r_s = (\frac{4}{3}\pi n_+ a_0^3)^{1/3} = 3$ to $0 < z/\text{Å} < 54$. A damping $\gamma = 0.025\ \omega_p$ is chosen ($\omega_p^2 = 4\pi e^2 n_+/m$). /4.32/

There has also been a considerable amount of work giving formal prescriptions how to calculate electromagnetic fields from a given nonlocal susceptibility, or deriving formal expressions for surface response functions. Some of the latter will be discussed in Chap. 5.

4.5 Collective and Single-Particle Response in the SCIB Model

An interesting feature of the surface fields revealed by the microscopic RPA calculations /3.84 ; 4.25 , 30/ is the occurrence of Friedel-type oscillations well below the plasma frequency. These oscillations originate from optical excitation of electron-hole pairs, which is forbidden in the bulk metal by the requirement of energy and momentum conservation. By the presence of the surface, translational invariance is broken and momentum is not a good quantum number, so that electron-hole excitations become possible. It has been emphasized by several authors /3.83, 84 ; 4.14/ that these effects are already included (at least qualitatively) in the phenomenological SCIB model, which uses Lindhard's RPA dielectric functions in the Kliewer-Fuchs integral (4.35, 36) for the electric field, whereas no Friedel-type oscillations can occur within the hydrodynamic model, which considers plasma oscillations as the only response modes of the electron system. Moreover, it was pointed out that within the SCIB model the contributions of collective plasmon excitation to the electric field can be separated analytically from the contributions giving rise to Friedel oscillations, if the Kliewer-Fuchs integrals are evaluated in the complex k plane. We now briefly report such an evaluation of the SCIB model /4.16/, which elucidates some relations between the SCIB model, the hydrodynamic model, and microscopic models.

We first recall some properties of Lindhard's RPA dielectric functions /4.35/, which can be obtained by evaluating (4.44) for the homogeneous bulk jellium model. With the notations

$$Q = k/k_F \quad , \quad \Omega = \hbar(\omega + i0^+)/E_F \quad , \quad \Omega_p = \hbar\omega_p/E_F \quad , \tag{4.54}$$

where $\omega_p = (4\pi e^2 n/m)^{1/2}$ is the plasma frequency, Lindhard's longitudinal and transversal dielectric functions can be written as /2.3 ; 4.16 , 36 , 37/:

$$\varepsilon_\ell(k,\omega) = 1 + \frac{3\Omega_p^2}{8Q^2}\left\{1 + \frac{1}{8Q^3}\sum_{\sigma=\pm1} F_\sigma L_\sigma\right\} \tag{4.55}$$

and

$$\varepsilon_t(k,\omega) = 1 - \frac{3\Omega_p^2}{32\Omega^2}\left\{Q^2 + 3\frac{\Omega^2}{Q^2} + 4 + \frac{1}{8Q^5}\sum_{\sigma=\pm1} F_\sigma^2 L_\sigma\right\} \quad , \tag{4.56}$$

respectively. Here $F_\sigma = (Q^2 - \sigma\Omega)^2 - 4Q^2$,

$$L_\sigma = Ln(Q + a_\sigma) + Ln(Q - b_\sigma) - Ln(Q + b_\sigma) - Ln(Q - a_\sigma) \quad , \tag{4.57}$$

where Ln(z) denotes the principal branch of the logarithm,

$$Ln(z) = \ln|z| + i\,arg(z) \quad \text{for} \quad -\pi < arg(z) \leqslant \pi \quad , \tag{4.58}$$

and $a_\sigma = (1 + \sigma\Omega)^{1/2} - 1$ with $Im\{a_\sigma\} > 0$, and $b_\sigma = a_\sigma + 2$. Both ε_ℓ and ε_t have branch cuts in the complex Q plane, terminating at the points $\pm a_\sigma$, $\pm b_\sigma$. If the cuts are chosen suitably /4.16/, ε_ℓ has a single zero in the upper halfplane $Im|Q| > 0$ (on the imaginary Q axis for $\omega < \omega_p$). The location of this zero and the endpoints of the cuts are shown in Fig. 4.2. (The constant value $Re\{b_-\} = -Re\{a_-\} = +1$ for $\omega > \omega_F$ is not indicated). For real k, both ε_ℓ and ε_t are real outside the particle-hole continuum, i.e. for $|Q| < a_+$ and $|Q| > b_+$. Figs. 4.3 and 4.4 give an idea of the k dependence of the RPA dielectric functions. For small and also for large real values of k simple expansions hold,

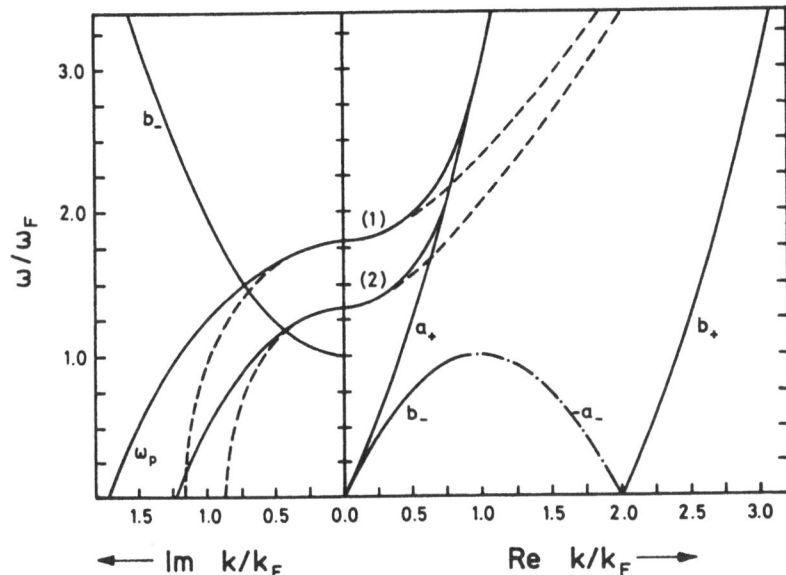

Fig. 4.2. Location of branch cuts and plasmon pole [zero of $\varepsilon_\ell(k,\omega)$] in the half-plane $\mathrm{Im}|k| \geqslant 0$. For a given value of ω, a_+ and b_+ terminate a cut along the real axis. The endpoints a_- and b_- of the second cut move for $\omega > \omega_F$ into the upper halfplane (then $\mathrm{Re}\{b_-\} = -\mathrm{Re}\{a_-\} = 1$ and only $\mathrm{Im}\{b_-\} = \mathrm{Im}\{a_-\}$ is shown for $\omega > \omega_F$). The plasmon pole is indicated for $r_s = 3.627$ (1) and $r_s = 2$ (2) for the RPA (———) and the HD with $\gamma = 0^+$ (-----). Its location is either on the imaginary ($\omega < \omega_p$) or on the real ($\omega > \omega_p$) axis

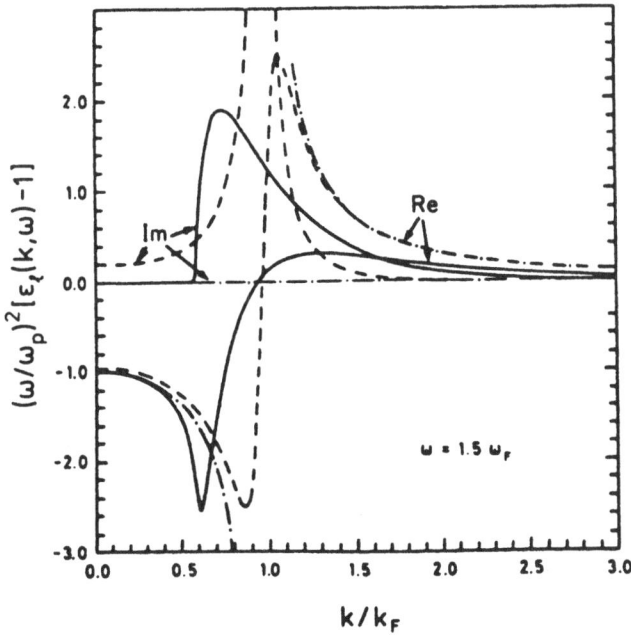

Fig. 4.3. k dependence of the longitudinal dielectric function in the RPA (———) and the HD with damping ($\gamma = 0.3\,\omega_F$, -----) and without damping ($\gamma = 0^+$, —·—·—, ε_ℓ real). Real and imaginary part of $(\omega/\omega_p)^2 [\varepsilon_\ell(k,\omega) - 1]$ are shown

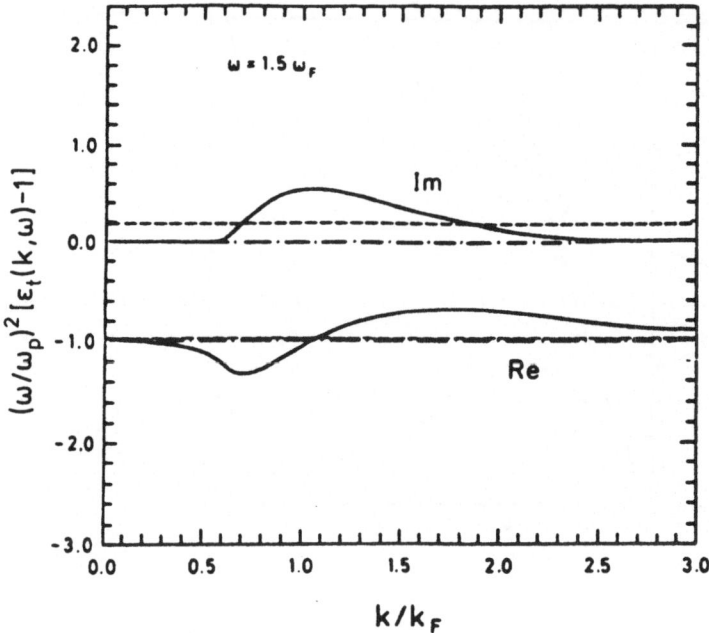

Fig. 4.4. k dependence of the transverse dielectric function in the RPA (———).
The constant values of the HD with ($\gamma = 0.3\ \omega_F$) and without ($\gamma = 0^+$) damping are
indicated by the *dotted* and the *dashed lines*, respectively. Again, $(\omega/\omega_p)^2$
$[\varepsilon_t(k,\omega) - 1]$ is shown

$$\varepsilon_\ell \approx 1 - \frac{\Omega_p^2}{\Omega^2}\left(1 + \frac{12}{5}\frac{Q^2}{\Omega^2}\right) \quad , \qquad \varepsilon_t \approx 1 - \frac{\Omega_p^2}{\Omega^2}\left(1 + \frac{4}{5}\frac{Q^2}{\Omega^2}\right) \tag{4.59}$$

if $|Q| \ll 1$, and $|Q| \ll \Omega$, and

$$\varepsilon_\ell \approx 1 + \frac{\Omega_p^2}{Q^4} \quad , \qquad \varepsilon_t \approx 1 - \frac{\Omega_p^2}{\Omega^2}\,(1 - \frac{4}{5Q^2}) \tag{4.60}$$

for $|Q| \gg 1, \Omega$. Note that $\varepsilon_t(\infty,\omega) = \varepsilon_t(0,\omega)$.

We now turn to the evaluation of the Kliewer-Fuchs integrals (4.35, 36, 42).
The transversal dielectric function occurs only in the combination $\varepsilon_t(k,\omega) - c^2 k^2/\omega^2$,
and the dominant contributions to the corresponding integrals come from k values
near the zero of this expression. Since $c \gg v_F$, this zero occurs for optical frequencies ($\hbar\omega$ and E_F of the same order of magnitude) at very small k values, $k \ll k_F$,
where $\varepsilon_t(k,\omega) \approx \varepsilon_t(0,\omega)$. For large k values, $k \gtrsim 0.1k_F$, $(ck/\omega)^2 \gg |\varepsilon_t(k,\omega)|$, and
the contribution to the integral rapidly decreases. As a consequence, it is an ex-
cellent approximation to neglect the spatial dispersion in ε_t within the SCIB model:
$\varepsilon_t(k,\omega) \equiv \varepsilon_t(0,\omega) = \varepsilon_\ell(0,\omega)$. Then the electric field, (4.35, 36), can be written as
$\mathbf{E} = \mathbf{E}^t + \mathbf{E}^\ell$ where the transverse field is given by the classical expression

$$E_x^t(z) = -\frac{i}{\pi} D_z(0) \int_{-\infty}^{+\infty} \frac{dk_z}{k^2} e^{izk_z} \left[\frac{k_z^2}{k_x(\varepsilon_t - c^2 k^2/\omega^2)} + \frac{k_x}{\varepsilon_t} \right]$$

$$\tag{4.61}$$

$$= -\frac{k_z^t}{k_x \varepsilon_t} D_z(0) e^{izk_z^t} \quad ,$$

$$E_z^t(z) = \frac{i}{\pi} D_z(0) \int_{-\infty}^{+\infty} \frac{dk_z}{k^2} e^{izk_z} \left[\frac{k_z}{\varepsilon_t - c^2 k^2/\omega^2} - \frac{k_z}{\varepsilon_t} \right]$$

$$\tag{4.62}$$

$$= \frac{1}{\varepsilon_t} D_z(0) e^{izk_z^t} \quad ,$$

with $k_x^2 + (k_z^t)^2 = \varepsilon_t \omega^2/c^2$, $\text{Im}\{k_z^t\} \geq 0^+$, $\text{Re}\{k_z^t\} \geq 0^+$, and the longitudinal field is given by

$$E_\mu^\ell(z) = \frac{i}{\pi} D_z(0) \int_{-\infty}^{+\infty} dk_z \, e^{izk_z} \frac{k_\mu}{k^2} \left[\frac{1}{\varepsilon_\ell(0,\omega)} - \frac{1}{\varepsilon_\ell(k,\omega)} \right]$$

$$\tag{4.63}$$

for $\mu = x,z$. Equation (4.37) yields $\mathbf{D} = \varepsilon_t \mathbf{E}^t$.

The integral (4.63) can be evaluated by the method of contour integrals in the complex plane /4.16/. As can be seen from Fig. 4.2 (from the intersection of the shown structures with the line ω = const.), for $\omega < \omega_p$ there arise, for example, an exponentially damped, collective plasmon contribution from the zero of $\varepsilon_\ell(k,\omega)$ and two oscillating Friedel-type contributions from the branch cuts terminating at a_+ and b_+ and at a_- and b_-, respectively. The oscillatory terms contain wave-vectors $k_F a_\pm$, $k_F b_\pm$, i.e. $[k_F^2 \pm 2m\hbar\omega]^{1/2} \pm k_F$ (if real), and decay as z^{-2}. They survive and become even enhanced in a microscopic RPA treatment of the surface /3.85/. A numerical evaluation and a detailed discussion of the respective role of pole and cut contributions has been given in /4.16/. Near the plasma frequency, $|\omega - \omega_p| \lesssim 0.1\,\omega_p$, the plasmon contribution clearly dominates the longitudinal electric field. Moreover, it was found that in the whole frequency regime $\omega \lesssim 1.3\,\omega_p$ the amplitudes of the cut contributions to the longitudinal electric field were always smaller than 20 % of the total variation of the longitudinal field. Figure 4.5 gives a typical example for the different contributions at a frequency well below the plasma frequency. According to a suggestion of FEIBELMAN /3.84 , 86/, the z component of the longitudinal field is given with the normalization (cf. Sect. 3.10):

$$\xi(z) = E_z^\ell(z)/[D_z(0)(\varepsilon_t^{-1} - 1)] = E_z^\ell(z)/[E_z^t(0^+) - D_z(0)] \quad .$$

$$\tag{4.64}$$

Then $\xi(0^+) = -1$ means that $E_z(z)$ is continuous at the surface, and $\text{Im}\{\xi(z)\}$ is closely related to the power absorption [cf. (4.71) below]. We note that careful

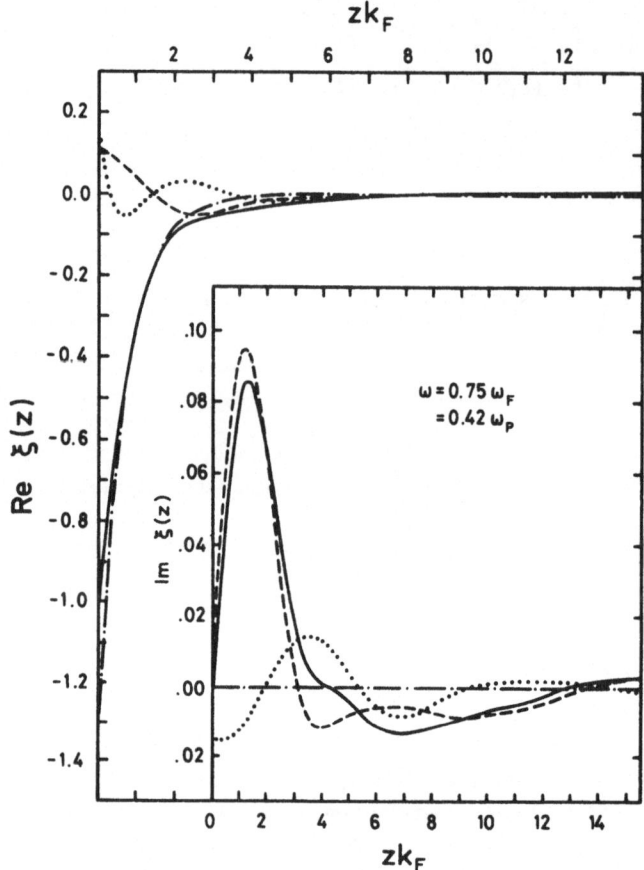

Fig. 4.5. Typical SCIB result for the normalized longitudinal electric field (4.64) (——), and its constituent contributions due to the plasmon pole (—·—·—), the cut (a_+, b_+) (— — —), and the cut (a_-, b_-) (·····) for $r_s = 3.627$ and $\omega < \omega_F$. $\xi(0^+) = -1$ means continuous $E_z(z)$. The plasmon contribution is real (since $\omega < \omega_p$), the cut contributions exhibit Friedel oscillations. For $\omega_F < \omega < \omega_p$ similar curves result, the cut (a_-, b_-) contributes, however, only an exponentially damped oscillation to $\mathrm{Re}\{\xi\}$ /4.16/

evaluation of all contributions is necessary to obtain the correct boundary value $\xi(0^+) = -1$. It is not a reasonable procedure to calculate the cut contributions exactly and to evaluate the plasmon pole only approximately, e.g. from the hydrodynamic model, as has been proposed in the literature /4.14, 15, 38/, since such inconsistencies lead to a discontinuous E_z /4.15/. Equation (4.40) for the correct boundary value of j_z may be considered as a sumrule which must be exhausted by the different contributions and should not be violated by inconsistent approximations.

Since the collective plasmon pole is found /4.16/ to yield an important contribution to the longitudinal field within the SCIB model, a quantitative comparison with the hydrodynamic model (HD), which considers only the response via plasmons, is of interest. We note that the dielectric function of the HD can be considered as an approximation of the RPA result. For $\omega \approx \omega_p$, the plasmon pole occurs at small k values where, according to (4.59),

$$\varepsilon_\ell \approx 1 - (\omega_p^2 + \beta k^2)/[\omega(\omega + i\gamma)] \quad , \quad \gamma = 0^+ \quad , \tag{4.65}$$

with $\beta = 3v_F^2/5$. This small-k expansion gives, however, no reasonable description of short-wavelength response and leads, if inserted into the Kliewer - Fuchs integrals (4.63), to the poor result that E_z and j_z are discontinuous at the surface [cf. (4.39)]. A much better result is obtained, if the RPA formula (4.55) is replaced by the hydrodynamic model

$$\varepsilon_\ell^{HD}(k,\omega) = 1 - \omega_p^2/[\omega(\omega + i\gamma) - \beta k^2] \quad , \quad \gamma = 0^+ \quad , \tag{4.66}$$

which has the same small-k behaviour (4.59) as ε_ℓ^{RPA}, but in addition satisfies [cf. (4.60)]

$$\varepsilon_\ell^{HD}(\infty,\omega) = \varepsilon_\ell^{RPA}(\infty,\omega) = 1 \quad . \tag{4.67}$$

Inserting ε_ℓ^{HD} together with $\varepsilon_t^{HD}(k,\omega) \equiv \varepsilon_\ell^{HD}(0,\omega)$ in the specular reflection formalism, we reobtain the results of the hydrodynamic model: E_z and j_z are continuous and no singular surface charges occur. We remark that within Sauter's partial-wave method, which was discussed in Sect. 2.4, there is no difference between the models (4.65) and (4.66), since both yield the same dispersion relation from $\varepsilon_\ell(k,\omega) = 0$. Within the Kliewer - Fuchs formalism, however, only (4.66) provides a useful model.

Longitudinal fields calculated /4.16/ for the SCIB model are in Fig. 4.6 compared with those calculated for the hydrodynamic model. Immediately at ω_p the agreement is excellent. Well below ω_p differences arise due to the cut contributions which in the SCIB model lead to Friedel-type oscillations and to related power absorption at the surface /3.84 ; 4.30/. Within the HD all response degrees of freedom are included in the collective plasmon mode, which exhausts the sumrule (4.40) and produces the correct boundary value $\xi(0^+) = -1$, but yields no absorption. In the HD absorption can be taken into account by a phenomenological damping constant γ, which then leads to the k dependence of the dielectric functions indicated in Figs. 4.3 , 4. A qualitative difference between HD and SCIB model remains, since $\gamma > 0$ introduces also power absorption due to transverse fields in the bulk, which is not included in the SCIB model. This is easily seen if the total power absorption (per unit surface area)

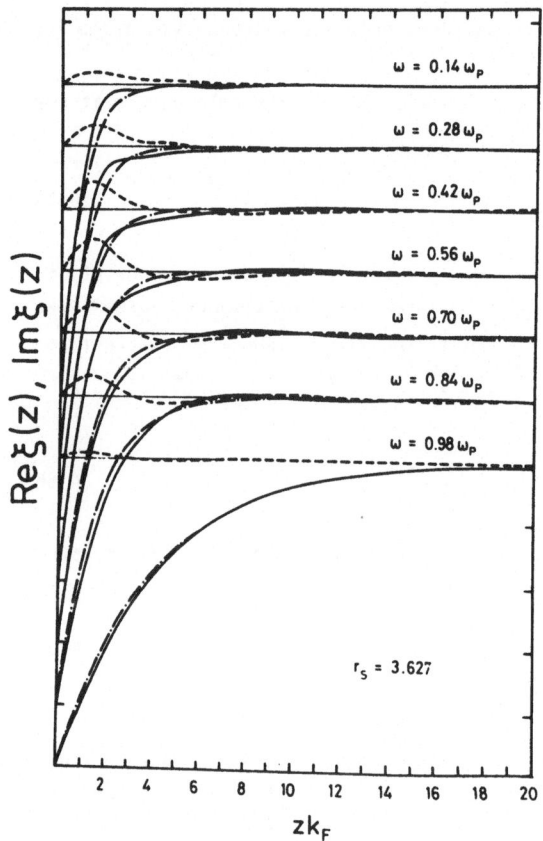

Fig. 4.6. Comparison of longitudinal fields $\xi(z)$ calculated for the SCIB and the hydrodynamic ($\gamma = 0^+$) model. (———,– – – –) give $\text{Re}|\xi|$ and $\text{Im}|\xi|$, respectively, for the SCIB model. For the HD (–·–·–) $\xi(z)$ is real for $\omega < \omega_p$. For each value of ω, $\text{Re}\{\xi(z)\}$ varies from $\xi(0^+) = -1$ at the surface to zero inside the metal ($z \to \infty$). $\text{Im}\{\zeta(z)\}$ is plotted on the same scale

$$P = \frac{1}{2} \int_0^\infty dz \ \text{Re}\{\mathbf{j}(z) \cdot \mathbf{E}(z)^*\} = -\frac{\omega}{8\pi} \int_0^\infty dz \ \text{Im}\{\mathbf{D}(z)^* \cdot \mathbf{E}(z)\} \qquad (4.68)$$

is [note that $\mathbf{j} = i\omega(\mathbf{E} - \mathbf{D})/4\pi$] calculated from (4.61 - 63). Using $\mathbf{D} = \varepsilon_t \mathbf{E}^t$, one obtains $P = P_B + P_S$ with a "bulk term"

$$P_B = \frac{\omega}{8\pi} \ \text{Im} \left\{ -\frac{1}{\varepsilon_t} \right\} \int_0^\infty dz \ |\mathbf{D}(z)|^2 \qquad (4.69)$$

including only the transverse field, and a "surface term"

$$P_S = -\frac{\omega}{8\pi} \int_0^\infty dz \ \text{Im}\{\mathbf{D}(z)^* \cdot \mathbf{E}^\ell(z)\}$$

$$= \frac{\omega}{8\pi^2} |D_z(0)|^2 \int_{-\infty}^{+\infty} \frac{dk_z}{k^2} \ \text{Im}\left\{ \frac{1}{\varepsilon_\ell(0,\omega)} - \frac{1}{\varepsilon_\ell(k,\omega)} \right\} \qquad , \qquad (4.70)$$

which depends explicitly on the longitudinal field. For the SCIB model, $\varepsilon_t = \varepsilon_\ell(0,\omega)$ is real, P_B is zero and, for $\omega < \omega_p$, only the particle-hole continuum contributes to P_S. The hydrodynamic model with $\gamma = 0^+$, on the other hand, yields, according to (4.68 - 70), no power absorption for $\omega < \omega_p$. Inserting $\gamma > 0$ into (4.66), one obtains, with $\varepsilon_t = \varepsilon_\ell(0,\omega)$, $P_B > 0$ but $P_S < 0$. The total absorption $P = P_B + P_S$ is, of course, positive within the HD, as was discussed in Sect. 2.5. A surface specific absorption mechanism, owing directly to the breaking of translational symmetry, is, however, not contained in the hydrodynamic model.

If bulk damping is included, as for instance in the HD or phenomenological exten-sions of the RPA /4.25/, the separation of surface absorption from bulk absorption becomes somewhat arbitrary /3.84/, since both transverse and longitudinal fields contribute to the absorption near the surface. Then the separation (4.69 , 70), which in a simple manner takes care of the latent infrared divergency of the integral in P_S owing to $Im\{-1/\varepsilon_\ell(k \to 0,\omega)\} > 0$, may need modification /4.26/ in order to have a positive definite "surface absorption". A clear definition of a local absorption density has been worked out, as far as we know, only within the hydrodynamic approx-imation and is discussed in Sect. 2.5.

The comparison of SCIB model and HD seems to favour the former, since it includes the single-particle response modes in addition to the collective modes. However, this improvement over the HD becomes marginal, if the resulting longitudinal fields are compared with those from microscopic calculations. Microscopic results for $Re\{\xi(z)\}$ exhibit much larger oscillations /3.84 ; 4.30/ than the corresponding SCIB results, whereas the surface power absorption can be much larger /3.84/ or much smaller /4.30 , 32/ than the SCIB absorption, depending on details of the surface. In sufficiently diffuse surfaces, excitation of collective surface modes is possible, which has an important effect on the surface response properties, as is discussed in Sect. 3.10 and also in Sect. 5.6. A first understanding of these effects was achieved by hydrodynamic model calculations including a surface layer of reduced electron density /3.32 , 33/. The specular reflection model, on the other hand, can-not include effects owing to surface diffuseness, and tractable generalizations of the SCIB model which include such effects are not known.

Before we close the chapter, we want to present the contribution (4.70) of the longitudinal field to the power absorption in a more convenient form. In the long-wavelength limit we may neglect E_x^ℓ (being much smaller than E_z^ℓ) and replace $D_z(z)$ by $D_z(0)$ to obtain with (4.64)

$$P_S \approx -\frac{\omega}{8\pi} \int_0^\infty dz \; Im\{D_z(0) \cdot E_z^\ell(z)\} = \frac{\omega}{8\pi} |D_z(0)|^2 \int_0^\infty dz \; Im\{(1 - \frac{1}{\varepsilon_t}) \xi(z)\} \quad . \quad (4.71)$$

Dividing by the incident flux $\Phi = (c/8\pi)|E_0|^2 \cos\theta$, where θ is the angle of inci-dence and E_0 the amplitude of the incident light wave, and expressing $D_z(0) =$

$(1 + r_p)E_0 \sin\theta$ in terms of the reflection amplitude r_p (notation as in Sect. 5.2 below), we obtain the "surface absorptance"

$$Y = P_S/\Phi = \frac{\omega}{c} \frac{\sin^2\theta}{\cos\theta} |1 + r_p|^2 \text{Im}\{(\frac{1}{\varepsilon_t} - 1)d_\perp\}$$

(4.72)

in terms of the surface response function

$$d_\perp(\omega) = -\int_0^\infty dz \, \xi(z)$$

(4.73)

to be discussed in Chap. 5 [cf. (5.44)]. For $\omega < \sqrt{2}\omega_p$ and for $\theta = 45^\circ$ the prefactor in (4.72) is close to $\sqrt{8}(\omega/c)[(\omega/\omega_p)^2 - 1]^2$, the value which is usually taken in plots of Y (cf. Fig. 4.1).

5. Description of Nonlocal Effects by the Surface Response Functions $d_\perp(\omega)$ and $d_\parallel(\omega)$

5.1 Economical Presentation of Experimental Results: $d_\perp(\omega)$, $d_\parallel(\omega)$

Nonlocal effects in metal optics lead to rapidly varying longitudinal fields near the surface, but far from the surface only transverse electromagnetic fields survive. This is true even at and above the plasma frequency, since the damping of plasma waves is typically by a factor c/v_F larger than that of the transverse waves. Within the very successful classical Fresnel optics, which considers only transverse fields, all the optical properties of a clean metal surface are determined by the bulk dielectric function of the metal (and the adjacent medium), which is a function of frequency only. It seems desirable to have a similar description of the nonlocal surface effects in terms of one or two general functions which depend only on frequency and allow to calculate all the optical properties, e.g. surface plasmon dispersion or reflection amplitudes, which may depend also on the angle of incidence of the incoming light, for instance.

FEIBELMAN /3.84 , 86 ; 5.1/ has shown from microscopic considerations that this can indeed be achieved in the "long-wavelength limit" (LWL), i.e. if the scale of the spatial variation of the transverse electromagnetic fields is much larger than the width of the surface region, in which deviations from the asymptotic transverse fields are important. With typical metals the conditions for the LWL are met below and, if a realistic damping is taken into account, also at and even well above the plasma frequency, and the nonlocal surface effects on the reflection amplitude and other measurable quantities can be expressed in terms of two surface response functions $d_\perp(\omega)$ and $d_\parallel(\omega)$, which depend only on the frequency ω. From a microscopic point of view, these surface response functions involve integrals over the surface region, and they cannot be used to calculate the surface fields. But they can be used to calculate the macroscopic response properties of clean surfaces and also of surfaces covered with thin films, and they can, in turn, be evaluated from experimental results. They offer a general, meaningful and economical way to present experimental and also theoretical results on optical properties of metal surfaces.

The aim of this Chap. 5 is to introduce these surface response functions, to establish their relation to observable quantities, and to give an idea about the physical origin and the implications of their frequency dependence.

A very transparent method to derive the surface parameters $d_\perp(\omega)$ and $d_\parallel(\omega)$, which also clarifies their physical meaning, has been proposed by APELL /5.2/ and is presented in a slightly generalized form in Sect. 5.2. The idea is old /5.3 - 5/ and has already been introduced by PLIETH and NAEGELE /5.6/ in the present context: One extrapolates the asymptotic transverse fields towards the surface and derives boundary conditions for these fields by an integration of Maxwell's equations in the surface region. These boundary conditions contain certain moments of the deviations of the exact fields from the extrapolated fields, i.e. of the "surface solutions" discussed by MUKHOPADHYAY and LUNDQVIST /4.14/, and yield exact expressions for, e.g., the reflection coefficient in terms of these moments. In Sect. 5.3 we show that these expressions reduce in the LWL to Feibelman's results and to equivalent expressions given by BAGCHI et al. /5.7/.

In Sect. 5.4 we consider the simple local three layer model, which has been discussed by McINTYRE and ASPNES /5.8/ and is frequently used to present experimental data. In the LWL such a model can be used to express the surface parameters $d_\perp(\omega)$ and $d_\parallel(\omega)$ in terms of the thickness d of the surface layer and the dielectric constants of surface layer and metal substrate, provided the surface layer has a reduced symmetry ($\varepsilon_{xx}^S = \varepsilon_{yy}^S \neq \varepsilon_{zz}^S$). But it is not possible to determine these optical constants and the thickness d of the surface layer uniquely from the values of $d_\perp(\omega)$ and $d_\parallel(\omega)$, or from optical measurements, as has been emphasized by PLIETH and NAEGELE /5.6/. Moreover, the nonlocal calculation of Sect. 5.5 shows that it does in general not increase the insight into the physics of the problem, if one expresses the surface response functions $d_\perp(\omega)$ and $d_\parallel(\omega)$ in terms of parameters of a local model, even if this is formally possible. A simple example is illustrative: The decay length of plasma waves (for $\omega < \omega_p$) and, thereby, the effective width of the surface region depends on the frequency. To simulate this effect in a local three layer model one needs a surface layer with an artificial frequency dependence of either the layer thickness or the dielectric functions. Furthermore, it turns out that only within the LWL nonlocal effects can be simulated by a local three layer model, so that there is no good reason to express experimental data in terms of dielectric functions of such a model.

In Sect. 5.5 we consider within the hydrodynamic approximation a three layer model in which both the surface layer and the bulk metal can sustain longitudinal fields. Within the LWL we present explicit analytical results for $d_\perp(\omega)$, $d_\parallel(\omega)$ and for the ellipsometry parameters, which contain previous results of ABELES and LOPEZ-RIOS /5.9/ as special cases and may be useful for the interpretation of experimental data on metal films adsorbed on metallic substrates. In Sect. 5.6 we discuss surface plasmons in terms of the response functions $d_\perp(\omega)$, $d_\parallel(\omega)$. Especially the treatment of "multipole" surface plasmons yields some understanding of the frequency dependence of $d_\perp(\omega)$.

5.2 Boundary Conditions for the Asymptotic Fields

We assume that far from the surface the exact electromagnetic fields reduce to transverse fields and compare the exact solution $\mathbf{E}(\mathbf{r}) = \mathbf{E}(z)\exp[i(k_x x - \omega t)]$ of Maxwell's equations in the whole space with a reference field defined by

$$\mathbf{E}^0(z;a) = \mathbf{E}^<(z)\theta(a - z) + \mathbf{E}^>(z)\theta(z - a) \quad , \tag{5.1}$$

where the transverse fields $\mathbf{E}^<(z)$ and $\mathbf{E}^>(z)$ are the extrapolations of the asymptotic limits of $\mathbf{E}(z)$ on the vacuum side and on the metal side, respectively, towards a plane $z = a$ in the surface region. The reference field (5.1) together with the corresponding $\mathbf{B}^0$-field and the displacement field

$$\mathbf{D}^0(z;a) = \varepsilon_a \mathbf{E}^<(z)\theta(a - z) + \varepsilon_t \mathbf{E}^>(z)\theta(z - a) \tag{5.2}$$

is assumed to solve Maxwell's equations with the local dielectric constants ε_a in the halfspace $z < a$ and ε_t in the metallic halfspace $z > a$. Here and in the following we assume the *local approximation* $\varepsilon_t(k,\omega) = \varepsilon_t(0,\omega) = \varepsilon_t$ to be *sufficient* for the bulk response of the metal to *transverse waves*. Furthermore we consider the slightly more general case that to the left of the surface we have a dielectric described by ε_a, rather than vacuum.

Since the reference fields are determined by the asymptotic values of the exact field, the reference fields will in general not satisfy the standard matching conditions at the plane $z = a$. That means, the reference field is not the solution of the classical Fresnel problem with dielectric constants ε_a and ε_t in $z < a$ and $z > a$, respectively. On the contrary, the reference field contains by definition the full information about the reflection and transmission properties of the nonlocal surface problem. Following APELL /5.2/, and, more closely, recent work by KEMPA and GERHARDTS /3.52/ we now derive the exact matching conditions for the reference field.

To be specific, we consider first the case of *p polarization* and write the field in the dielectric in the form

$$E_x^<(z) = -\frac{cp_a}{\omega} E_0(e^{izp_a} - r_p e^{-izp_a}) \quad , \tag{5.3a}$$

$$E_z^<(z) = \frac{ck_x}{\omega} E_0(e^{izp_a} + r_p e^{-izp_a}) \quad , \tag{5.3b}$$

with E_0 the amplitude of the incident field, r_p the reflection amplitude and $k_x^2 + p_a^2 = \varepsilon_a \omega^2/c^2$. The asymptotic transverse field inside the metal is written as

$$E_x^>(z) = E_x^t e^{izp_t} \quad , \tag{5.4a}$$

$$E_z^>(z) = -\frac{k_x}{p_t} E_x^t e^{izp_t} \quad , \tag{5.4b}$$

with $k_x^2 + p_t^2 = \varepsilon_t \omega^2/c^2$. By construction of the reference field there exist z values $\xi_1 < a$ and $\xi_2 > a$ (slightly) outside the surface region, so that the exact fields agree practically with the reference fields, e.g. $\mathbf{D}(z) \approx \mathbf{D}^0(z;a)$, for $z \lesssim \xi_1$ and for $z \gtrsim \xi_2$. Since both the exact fields and the reference fields satisfy in the halfspace $z > a$ and $z < a$ Maxwell's equations, although with different constitutive equations, we can use $\mathbf{\nabla \cdot D} = ik_x D_x + D_z' = 0$ for both $\mathbf{D(r)}$ and $\mathbf{D}^0(\mathbf{r};a)$ to evaluate

$$\int_a^{\xi_2} dz \, [D_z'(z) - D_z^{0'}(z;a)] = -[D_z(a) - D_z^>(a)] \tag{5.5}$$

$$= -ik_x \int_a^{\xi_2} dz \, [D_x(z) - D_x^0(z;a)] \quad ,$$

where $D_z(\xi_2) = D_z^0(\xi_2;a) = D_z^>(\xi_2)$ has been taken into account. Adding the corresponding integral over the interval $\xi_1 < z < a$, we obtain, since $D_z(z)$ is continuous at $z = a$,

$$D_z^>(a) - D_z^<(a) = -ik_x \int_{\xi_1}^{\xi_2} dz[D_x(z) - D_x^0(z;a)] \quad . \tag{5.6}$$

This matching condition for the asymptotic fields replaces the standard boundary condition "$D_z(z)$ continuous".

A second matching condition for the asymptotic fields, corresponding to the standard boundary condition "$E_x(z)$ continuous", is obtained from Faraday's law $\mathbf{\nabla \times E} = -c^{-1}\partial\mathbf{B}/\partial t$, i.e. $ik_x E_z - E_x' = -i\omega B_y/c$, which yields for instance

$$\int_a^{\xi_2} dz \, E_x'(z) = E_x(\xi_2) - E_x(a)$$

$$= ik_x \int_a^{\xi_2} dz \, E_z(z) + i\frac{\omega}{c}[\xi_2 B_y(\xi_2) - aB_y(a^+)] + \frac{\omega^2}{c^2}\int_a^{\xi_2} dz \, zD_x(z) \quad . \tag{5.7}$$

Here we have integrated by parts, using $B_y' = i\omega D_x/c$, the x component of Ampère's law $\mathbf{\nabla \times H} = c^{-1}\partial\mathbf{D}/\partial t$. We now substract from (5.7) the corresponding expression for the reference fields and add the result to that obtained in the same way by integrating over the interval $\xi_1 \leqslant z < a$. This yields the matching condition

$$E_x^>(a) - E_x^<(a) = ik_x \int_{\xi_1}^{\xi_2} dz[E_z(z) - E_z^0(z;a)] + \frac{\omega^2}{c^2}\int_{\xi_1}^{\xi_2} dz(z - a)[D_x(z) - D_x^0(z;a)] \quad , \tag{5.8}$$

where the explicit boundary terms at $z = \xi_1$ and $z = \xi_2$ have cancelled according to the definition of the reference fields. The boundary terms $\sim a[B_y^>(a) - B_y^<(a)]$ have been included in the last integral of (5.8) using $ik_x B_y = -i\omega D_z/c$ and (5.6).

For a compact notation we define the following moments of "surface solutions" /4.14/, i.e. of differences between exact and reference fields:

$$\delta_\mu^{(n)}(a) = \int_{-\infty}^{+\infty} dz \ (z-a)^{n-1}[D_\mu(z) - D_\mu^0(z;a)]/E_\mu^>(a) \quad , \tag{5.9a}$$

$$\eta_\mu^{(n)}(a) = \int_{-\infty}^{+\infty} dz \ (z-a)^{n-1}[E_\mu(z) - E_\mu^0(z;a)]/D_\mu^>(a) \quad , \tag{5.9b}$$

with $n = 1,2$ and $\mu = x,y$ or z. By construction only the surface region $\xi_1 < z < \xi_2$ contributes to the integrals, and the order of magnitude of the moments $\delta_\mu^{(n)}(a)$ and $\eta_\mu^{(n)}(a)$ can be estimated by $(\xi_2 - \xi_1)^n$.

With these moments and (5.2,4), we can rewrite the boundary conditions for the asymptotic fields at the reference plane $z = a$ in the case of p polarization as

$$D_z^>(a) - D_z^<(a) = \alpha_p(a) \ D_z^>(a) \quad , \tag{5.10a}$$

$$E_x^>(a) - E_x^<(a) = \beta_p(a) \ E_x^>(a) \quad , \tag{5.10b}$$

where

$$\alpha_p(a) = i \frac{p_t}{\varepsilon_t} \delta_x^{(1)}(a) \quad , \tag{5.11a}$$

$$\beta_p(a) = -i \frac{k_x^2}{p_t} \varepsilon_t \ \eta_z^{(1)}(a) + \frac{\omega^2}{c^2} \delta_x^{(2)}(a) \quad . \tag{5.11b}$$

Using (5.2) to (5.4) and (5.10), one easily solves for the reflection amplitude,

$$r_p = \frac{\varepsilon_t p_a(1-\alpha_p) - \varepsilon_a p_t(1-\beta_p)}{\varepsilon_t p_a(1-\alpha_p) + \varepsilon_a p_t(1-\beta_p)} \ e^{2iap_a} \quad . \tag{5.12}$$

The transmission amplitude, defined as the ratio of z components of transmitted and incident field, $t_p = E_z^>(0)/(ck_x E_0/\omega)$, is obtained as

$$t_p = \frac{2\varepsilon_a p_a}{\varepsilon_t p_a(1-\alpha_p) + \varepsilon_a p_t(1-\beta_p)} \ e^{ia(p_a-p_t)} \quad . \tag{5.13}$$

Whereas the moments $\delta_x^{(n)}(a)$ and $\eta_z^{(1)}(a)$ depend, as indicated, on the position of the reference plane $z = a$, all measurable quantities, e.g. r_p and t_p, must be independent of the particular choice of a. This becomes important if approximate formulas for the moments are used as, e.g., in the LWL discussed in Sect. 5.3.

If we neglect the nonlocal effects and replace the exact field by the reference field, then $\delta_x^{(n)}(a) = \eta_z^{(1)}(a) = 0$ and $\alpha_p(a) = \beta_p(a) = 0$. In this case (5.10) reduces to the standard matching conditions of the classical Fresnel theory for reflection at a surface plane at $z = a$, and (5.12) and (5.13) reduce to the classical values $r_p^{cl}(a)$ and $t_p^{cl}(a)$, respectively.

Equations (5.9 - 13) are an exact consequence of Maxwell's equations. If we neglect $\delta_x^{(2)}$ and put $a = 0$, we obtain the approximate result of APELL /5.2/, which is sufficient in long-wavelength limit (cf. Sect. 5.3).

The case of s *polarization* can be treated in a similar way. Instead of (5.3 , 4) we have

$$E_y^<(z) = E_0(e^{izp_a} - r_s e^{-izp_a}) \quad , \tag{5.14}$$

$$E_y^>(z) = E_t e^{izp_t} \quad . \tag{5.15}$$

Since $\mathbf{\nabla} \times \mathbf{E} = -c^{-1}\partial \mathbf{B}/\partial t$ now yields $E_y' = -i\omega B_x/c$ and $k_x E_y = \omega B_z/c$, we can integrate $\mathbf{\nabla} \times \mathbf{B} = c^{-1}\partial \mathbf{D}/\partial t$, i.e. $-ik_x B_z + B_x' = -i\omega D_y/c$, for the exact and the reference fields in the same way as we did above to obtain

$$B_x^>(a) - B_x^<(a) = -i\frac{\omega}{c}\delta_y^{(1)}(a)E_y^>(a) + i\frac{c}{\omega}k_x^2 \varepsilon_t \eta_y^{(1)}(a)E_y^>(a) \quad , \tag{5.16a}$$

which replaces the classical matching condition "B_x continuous". From $E_y' = -i\omega B_x/c$ we obtain the generalization of "E_y continuous",

$$E_y^>(a) - E_y^<(a) = \frac{\omega^2}{c^2}\delta_y^{(2)}(a)E_y^>(a) - k_x^2 \varepsilon_t \eta_y^{(2)}(a)E_y^>(a) \quad , \tag{5.16b}$$

where we have integrated by parts to eliminate B_x in favour of B_x', and we have used (5.9) and (5.16a). With the abbreviations

$$\alpha_s(a) = \frac{i}{p_t}\left[\frac{\omega^2}{c^2}\delta_y^{(1)}(a) - \varepsilon_t k_x^2 \eta_y^{(1)}(a)\right] \quad , \tag{5.17a}$$

$$\beta_s(a) = \frac{\omega^2}{c^2}\delta_y^{(2)}(a) - \varepsilon_t k_x^2 \eta_y^{(2)}(a) \quad , \tag{5.17b}$$

the generalized boundary conditions for the asymptotic fields at the reference plane $z = a$ can thus be written in the case of s polarization

$$B_x^>(a) - B_x^<(a) = \alpha_s(a)B_x^>(a) \quad , \tag{5.18a}$$

$$E_y^>(a) - E_y^<(a) = \beta_s(a)E_y^>(a) \quad . \tag{5.18b}$$

For the reflection amplitude r_s and the transmission amplitude $t_s = E_t/E_0$, we obtain from (5.14 , 15 , 18)

$$r_s = \frac{p_t(1-\alpha_s) - p_a(1-\beta_s)}{p_t(1-\alpha_s) + p_a(1-\beta_s)} e^{2iap_a} \quad , \tag{5.19}$$

$$t_s = \frac{2p_a}{p_t(1-\alpha_s) + p_a(1-\beta_s)} e^{ia(p_a-p_t)} \quad . \tag{5.20}$$

This again is an exact consequence of Maxwell's equations. The amplitudes r_s and t_s are independent of the position $z = a$ of the reference plane, whereas α_s and β_s depend on a. If we neglect nonlocal effects, and take $\delta_y^{(n)} = \eta_y^{(n)} = \alpha_s = \beta_s = 0$, (5.18a,b) reduce to the standard matching conditions and (5.19) and (5.20) to the classical values $r_s^{cl}(a)$ and $t_s^{cl}(a)$ of the Fresnel problem with a metal surface at $z = a$. Equations (5.12 , 13 , 19 , 20) give simple generalizations of the corresponding classical formulas for reflection and transmission amplitudes, which take full account of nonlocal effects at the surface through the moments $\delta_\mu^{(n)}$, $\eta_\mu^{(n)}$ defined in (5.9). The integrals in (5.9) are extended from $-\infty$ to $+\infty$, so that the boundaries ξ_1 and ξ_2 of the surface region, in which the difference between actual and reference fields is nonzero, do not appear explicitly. This form is also applicable for frequencies above the plasma frequency, where in the case of p polarization bulk plasmons can be excited, which lead to longitudinal field contributions extending deep into the metal.

In the most general case, the four parameters α_p , β_p , α_s , β_s, which determine the generalized boundary conditions (5.10 , 18) are complex functions of the frequency of the incident light and also of the angle of incidence, and their calculation requires the knowledge of the exact surface fields, as we demonstrate for two examples in Sects. 5.4 , 5. Then the concept of generalized boundary conditions for the asymptotic transverse fields is not useful, since it is of no help for the calculation of these fields. According to (5.12 , 13 , 19 , 20) the information contained in α , β is exactly the same as that contained in r , t, and there is no apparent reason to evaluate experiments in terms of α , β instead of r , t.

But the situation is different for typical metals and optical frequencies. Then the long-wavelength limit, to be discussed in the following section, is applicable, the angular dependence of the parameters α and β becomes trivial, and two complex functions of frequency, $d_\perp(\omega)$ and $d_\parallel(\omega)$, are sufficient to determine the four complex parameters α_p , β_p and α_s , β_s. In the LWL the generalized boundary conditions are useful to calculate angular dependencies, once the surface response functions $d_\perp(\omega)$ and $d_\parallel(\omega)$ are known. It is also reasonable to evaluate experimental results in terms of these surface response functions in order to eliminate trivial angular dependencies.

5.3 The Long-Wavelength Limit

In this section we show that in the LWL the surface response properties can be expressed in terms of two functions $d_\perp(\omega)$ and $d_\parallel(\omega)$, which are related to mean values of the nonlocal dielectric tensor and its inverse. For optical frequencies, the wavelength (or decay length) of the transverse field both inside and outside the metal is much longer than the typical length scale on which longitudinal fields and induced charge densities (and, of course, the unperturbed charge density) at the surface vary. Even at and above the plasma frequency the longitudinal fields decay (due to damping effects) much faster than transverse fields. This is easily visualized within the hydrodynamic approximation of Sect. 2.7, which yields for the z component of the wavevector of the longitudinal field, p_ℓ, and of the transverse field, p_t, (2.25 , 27),

$$p_\ell^2 = \frac{5}{3v_F^2} \left[\omega(\omega + i\gamma) - \frac{\omega_n^2}{\varepsilon_b} \right] - k_x^2 \qquad (5.21a)$$

and

$$p_t^2 = \frac{\omega \varepsilon_b}{\omega + i\gamma} \frac{1}{c^2} \left[\omega(\omega + i\gamma) - \frac{\omega_n^2}{\varepsilon_b} \right] - k_x^2 \quad , \qquad (5.21b)$$

respectively. Realistic values of the damping satisfy $(v_F/c)^2 \omega_p \ll \gamma \ll \omega_p$ and the decay length of the plasma waves is typically by a factor $v_F/c \sim 10^{-2}$ smaller than the decay length or wavelength of transverse waves, even at the plasma frequency ω_p. For typical s-p metals the plasma frequency is of the order of magnitude $\hbar\omega_p \sim$ 10 eV or less. For frequencies in this range (e.g. $\omega < 1.5\ \omega_p$) the transverse fields inside the metal vary only on a scale $\gtrsim 10^3$ Å. Here and in the following we assume, if not explicitly state otherwise, that the effective thickness $\tilde{d}$ of the surface region in which deviations from the asymptotic transverse field occur is of the order of several or, at most, several tens of Ångstroms, i.e. we consider clean surfaces or surfaces covered with thin films (width $\lesssim 10$ Å) and assume for $\omega > \omega_p$ a realistic damping of plasma waves. Then $\tilde{d}\omega/c \ll 1$, and in the rest of this Sect. 5.3 we retain only leading order terms with respect to the small parameter $\tilde{d}\omega/c$, i.e. we consider the long-wavelength limit (LWL), which has extensively been discussed by FEIBELMAN /3.84 - 86/.

In the LWL, simplifications of the following type occur. Field components, which are already continuous in the local Fresnel treatment, vary slowly in the surface region and are essentially constant over a distance $\tilde{d}$. Since the bulk response to transverse fields is assumed to be local anyhow, these field components can be taken out of the integrals which specify the nonlocal constitutive equations, so that these effectively reduce to local equations even in the surface region. Arguments of this type have extensively been discussed in the literature /3.84 ; 4.14 , 30 ; 5.2 , 7 , 10/ and will now be used to simplify the results of Sect. 5.2.

We consider first the simpler case of s *polarization*. In the LWL we may consider $E_y(z)$ as constant in the surface region, which implies [cf. discussion below (5.4)]

$$E_y(\xi_2) - E_y(\xi_1) = E_y^>(a) - E_y^<(a) = 0 \quad , \tag{5.22}$$

so that $E_y^0(z;a) = E_y(z)$ and $n_y^{(n)}(a) = 0$ from (5.9b). Since the response in the bulk, where only transverse fields survive, is assumed local and since $E_y(z)$ is practical-ly constant in the surface region, where nonlocal effects can occur, the constitu-tive equation

$$D_y(z) = \int_{-\infty}^{+\infty} dz' \; \varepsilon_{yy}(z,z') E_y(z') \approx \int_{-\infty}^{+\infty} dz' \; \varepsilon_{yy}(z,z') E_y(z) \tag{5.23}$$

reduces effectively to a local one. The effective local dielectric function

$$\varepsilon_{yy}(z) = \int_{-\infty}^{+\infty} dz' \; \varepsilon_{yy}(z,z') \tag{5.24}$$

interpolates smoothly between the values ε_a for $z < \xi_1$ and ε_t for $z > \xi_2$. As a con-sequence of (5.22 - 24), we obtain for (5.9a)

$$\delta_y^{(n)}(a) = \int_{-\infty}^{+\infty} dz \; (z - a)^{n-1} \left\{ \varepsilon_{yy}(z) - [\varepsilon_a \theta(a - z) + \varepsilon_t \theta(z - a)] \right\} \quad , \tag{5.25}$$

where, for $\xi_1 < a < \xi_2$, contributions to the integral come from $\xi_1 < z < \xi_2$. With $\tilde{d} = \xi_2 - \xi_1$ we estimate the order of magnitude $\delta_y^{(n)} \sim \tilde{d}^n$ $(n = 1,2)$, so that in the LWL only $\delta_y^{(1)}$ contributes to (5.18 - 20), whereas $|p_t \delta_y^{(2)}| \ll |\delta_y^{(1)}|$. Expanding the phase factors in (5.18 - 20) we obtain in the LWL ($|a| \lesssim \tilde{d}$), using $p_t^2 - p_a^2 = (\varepsilon_t - \varepsilon_a)\omega^2/c^2$,

$$r_s = \frac{p_t - p_a}{p_t + p_a} [1 + 2ip_a \; d_\parallel(\omega)] \quad , \tag{5.26}$$

$$t_s = \frac{2p_a}{p_t + p_a} [1 - i(p_t - p_a) \; d_\parallel(\omega)] \quad , \tag{5.27}$$

where

$$d_\parallel(\omega) = \frac{\delta_y^{(1)}(a)}{\varepsilon_a - \varepsilon_t} + a \quad . \tag{5.28}$$

For $a = 0$, (5.26) reduces to a result of BAGCHI et al. /5.7/, who used the notation $\Lambda_y(\omega)$ for $\delta_y^{(1)}(0)$. The surface function $d_\parallel(\omega)$, (5.28), is independent of a, as is easily seen from the derivative of (5.25) with respect to a. It has been introduced (for $\varepsilon_a = 1$) previously by FEIBELMAN and may be written in several equivalent forms /3.84 ; 5.1, 2/. Integrating by parts one obtains from (5.25), for instance,

$$d_{\parallel}(\omega) = \frac{1}{\varepsilon_t - \varepsilon_a} \int\limits_{-\infty}^{+\infty} dz \; z \frac{d}{dz} \varepsilon_{yy}(z) \quad . \tag{5.29}$$

We want to emphasize that the assumption $E_y(z)$ = const. and the neglect of $n_y^{(n)}$ is correct in lowest order of $\tilde{\omega d}/c$. This can be seen as follows. Using $k_x^2 n_y^{(2)} \ll \omega^2 \delta_y^{(2)}/c^2$ as input, we obtain from the generalized matching condition (5.18)

$$[E_y^>(a) - E_y^<(a)]/E_y^>(0) = \omega^2 \delta_y^{(2)}/c^2 \sim (\tilde{\omega d}/c)^2 \quad ,$$

instead of (5.22). This yields the estimate $[E_y(z) - E_y^0(z;a)]/E_y^>(0) \sim (\tilde{\omega d}/c)^2$ for $\xi_1 < z < \xi_2$ and, from (5.9b), $n_y^{(n)} \sim \tilde{d}^n(\tilde{\omega d}/c)^2$. Since $|k_x \tilde{d}| \ll 1$, the leading order results remain unaffected. In Sect. 5.4 we give an example which illustrates this in detail.

For *p polarization* the situation is a little more complicated, since in principle the dielectric tensor $\varepsilon_{\mu\nu}(q_x;z,z')$ is not diagonal, e.g., $\varepsilon_{xz} \neq 0$, and x and z components of electric and displacement field are coupled. But in the LWL this coupling is weak and can be neglected, as has been shown explicitly in a recent RPA calculation /4.30/, and as is plausible from the following argument. Since the isotropic transverse dielectric constant ε_t correctly describes the response deep inside the metal, the coupling can occur only in the surface region $(z,z' \sim \tilde{d})$. From rotational invariance of the metal around the surface normal one can easily show that ε_{xz} and ε_{zx} must be proportional to q_x. Then the nondiagonal elements of the dielectric tensor must be much smaller, typically by a factor of the order $q_x \tilde{d} \ll 1$, than the diagonal elements. By similar arguments /4.20/ one can show that in the LWL ε_{xx} equals ε_{yy}. The reason for these simplifications to hold is that q_x is much smaller than typical electronic momenta, $q_x \ll k_F$. (This should not be confused with the trivial limit $q_x \ll \omega/c$, i.e. nearly perpendicular incidence, which is neither required nor implied.)

Assuming that the x components of the E and D field are decoupled from the z components and that $\varepsilon_{xx} = \varepsilon_{yy}$, we proceed as above to obtain $\delta_x^{(1)}(a) = \delta_y^{(1)}(a)$ [cf. (5.25)] and $\delta_x^{(2)} = 0$ in the LWL. The same type of arguments cannot immediately be applied to the material equation

$$D_z(z) = \int\limits_{-\infty}^{+\infty} dz' \; \varepsilon_{zz}(z,z')E_z(z') \quad , \tag{5.30}$$

since here $D_z(z)$ is the slowly varying field, not $E_z(z)$ which, in the classical theory, is discontinuous at the surface. But since there is a one to one correspondence between **E** and **D** fields of Maxwell's theory, one can formally introduce the inverse dielectric tensor and replace (5.30) by the equation

$$E_z(z) = \int\limits_{\infty}^{+\infty} dz' \; \varepsilon_{zz}^{-1}(z,z')D_z(z') \quad , \tag{5.31}$$

which can be treated in the same way as (5.23): The slowly varying $D_z(z')$ is replaced with $D_z(z)$, an effective local function

$$\varepsilon_{zz}^{-1}(z) = \int_{-\infty}^{+\infty} dz' \ \varepsilon_{zz}^{-1}(z,z') \tag{5.32}$$

is introduced which interpolates smoothly between the values $1/\varepsilon_a$ and $1/\varepsilon_t$ deep in
the dielectric (vacuum) and the metal, respectively, and (5.9b) yields

$$\eta_z^{(1)}(a) = \int_{-\infty}^{+\infty} dz \left\{ \varepsilon_{zz}^{-1}(z) - \left[\frac{1}{\varepsilon_a} \theta(a - z) + \frac{1}{\varepsilon_t} \theta(z - a) \right] \right\} \quad . \tag{5.33}$$

Taking the derivative of $\eta_z^{(1)}$ with respect to a, we see that

$$d_\perp(\omega) = [\varepsilon_a^{-1} - \varepsilon_t^{-1}]^{-1} \ \eta_z^{(1)}(a) + a \tag{5.34}$$

is independent of the position a of our reference plane. This is (for $\varepsilon_a = 1$) the
second surface response function introduced by FEIBELMAN /3.84 , 86/. It may be written as

$$d_\perp(\omega) = \left(\frac{1}{\varepsilon_a} - \frac{1}{\varepsilon_t} \right)^{-1} \int_{-\infty}^{+\infty} dz \left\{ \varepsilon_{zz}^{-1}(z) - \left[\frac{1}{\varepsilon_a} \theta(-z) + \frac{1}{\varepsilon_t} \theta(z) \right] \right\} \quad , \tag{5.35}$$

and it determines together with

$$d_\parallel(\omega) = (\varepsilon_a - \varepsilon_t)^{-1} \int_{-\infty}^{+\infty} dz \left\{ \varepsilon_{xx}(z) - \left[\varepsilon_a \theta(-z) + \varepsilon_t \theta(z) \right] \right\} \tag{5.36}$$

the surface response properties in the long-wavelength limit [where $\varepsilon_{xx}(z) = \varepsilon_{yy}(z)$].
The reflection and transmission amplitudes given by (5.11 - 13) reduce in the LWL to

$$r_p = \frac{\varepsilon_t p_a - \varepsilon_a p_t - i(\varepsilon_a - \varepsilon_t)(p_a p_t d_\parallel - k_x^2 d_\perp)}{\varepsilon_t p_a + \varepsilon_a p_t - i(\varepsilon_a - \varepsilon_t)(p_a p_t d_\parallel + k_x^2 d_\perp)} \tag{5.37}$$

and

$$t_p = \frac{2 \varepsilon_a p_a}{\varepsilon_t p_a + \varepsilon_a p_t - i(\varepsilon_a - \varepsilon_t)(p_a p_t d_\parallel + k_x^2 d_\perp)} \quad , \tag{5.38}$$

respectively. To obtain these results we expand the exponentials in (5.12 , 13) to
first order in $p_a a$ and $(p_a - p_t)a$, respectively, and use the identity $(\varepsilon_t - \varepsilon_a) \cdot$
$(p_a p_t - \tau k_x^2) = (p_t + \tau p_a) \cdot (\varepsilon_t p_a - \tau \varepsilon_a p_t)$, which holds for $\tau = \pm 1$. As it should be,
the leading order results are independent of a. Equation (5.37) is useful for the
discussion of surface plasmons, as will be shown in Sect. 5.6. We will see that
$d_\perp(\omega)$ may become large, even within the LWL, if a large electric field is induced
in the surface region. This is possible under certain conditions and will be dis-

cussed in Sect. 5.6 together with the so called multipole surface plasmons. Apart from this interesting situation, $d_\perp$ will be small, $|d_\parallel|$, $|d_\perp| \sim \tilde{d} \ll c/\omega$, and we may expand the denominator of (5.37) to obtain

$$r_p = \frac{\varepsilon_t p_a - \varepsilon_a p_t}{\varepsilon_t p_a + \varepsilon_a p_t} \left\{ 1 + 2 i p_a \frac{\varepsilon_a p_t^2 d_\parallel - \varepsilon_t k_x^2 d_\perp}{\varepsilon_a p_t^2 - \varepsilon_t k_x^2} \right\} \tag{5.39}$$

where $\varepsilon_t^2 p_a^2 - \varepsilon_a^2 p_t^2 = (\varepsilon_t - \varepsilon_a)(\varepsilon_a p_t^2 - \varepsilon_t k_x^2)$, an immediate consequence of the identity

$$(p_a^2 + k_x^2)/\varepsilon_a = (p_t^2 + k_x^2)/\varepsilon_t = \omega^2/c^2 \quad ,$$

has been taken into account. Equation (5.39) agrees (for $\varepsilon_a = 1$) with FEIBELMAN's result /3.84/. Equivalent results have also been published by BAGCHI et al. /5.7/, who express r_p in terms of $\Lambda_z = n_z^{(1)}(0)$ and $\Lambda_x = \delta_x^{(1)}(0)$, and also by ABELES and LOPEZ-RIOS /5.9/, who are mainly interested in the ratio $\rho = r_p/r_s$ measured in ellipsometry experiments.

From (5.26 , 39) one derives for the ellipsometry ratio

$$\rho = \rho^{cl} \left\{ 1 + \frac{2 i p_a}{(\varepsilon_a/\varepsilon_t) - p_a^2/k_x^2} [d_\perp(\omega) - d_\parallel(\omega)] \right\} \quad , \tag{5.40}$$

where ρ^{cl} is the classical value of ρ, obtained from (5.26 , 39) neglecting $d_\perp$ and $d_\parallel$. The information about the surface is contained in (5.40) through the quantity $d_\perp - d_\parallel$, whereas its prefactor depends on the bulk dielectric constant ε_t of the metal and, via $k_x = (\omega/c)\varepsilon_a^{1/2} \sin\theta$ and $p_a = (\omega/c)\varepsilon_a^{1/2} \cos\theta$, on the angle of incidence θ. ABELES and LOPEZ-RIOS /5.9/ have emphasized that this structure has two consequences for the information abtainable from ellipsometry. First, for given frequency, only one complex number related to surface properties, $d_\perp(\omega) - d_\parallel(\omega)$, can be extracted from experiment. Second, the bulk value $\varepsilon_t(\omega)$ can be determined from ellipsometric measurements at different angles, at least in principle /5.9/.

The suggestive representation (5.35 , 36) of the surface functions $d_\perp(\omega)$ and $d_\parallel(\omega)$ is usually not convenient for explicit calculations, since one usually doesn't know both the dielectric tensor $\varepsilon_{\mu\nu}(z,z')$ of the surface problem and its inverse.

As an example we consider a jellium surface in the RPA, which yields an explicit microscopic expression for $\varepsilon_{\mu\nu}(z,z')$, so that the evaluation of $d_\parallel(\omega)$ is straightforward. In the LWL the tensor components $\varepsilon_{xx}(k_x=0;z,z') = \varepsilon_{yy}(k_x=0;z,z') = [1 - 4\pi e^2 n(z)/(m\omega^2)]\delta(z - z')$ become local /3.84 ; 5.2/. Then (5.36) reduces for $\varepsilon_a = 1$ to

$$d_\parallel(\omega) = \int_{-\infty}^{+\infty} dz \, [\theta(z) - n(z)/n(\infty)] \quad , \tag{5.41}$$

which is real and independent of frequency. If the origin $z = 0$ is chosen at the jellium edge, charge neutrality implies $d_\parallel = 0$ /3.84 ; 4.30 ; 5.2/.

In order to calculate $d_\perp$ from (5.35, 32), one formally needs the inverse of the RPA kernel $\varepsilon_{\mu\nu}(z,z')$. The calculation of this inverse is much more complicated than the solution of Maxwell's equations for an incident plane wave. Thus, for practical computation (5.35) is less convenient than the formula

$$d_\perp(\omega) = \left(\frac{1}{\varepsilon_a} - \frac{1}{\varepsilon_t}\right)^{-1} \int\limits_{\xi_1}^{\xi_2} dz \left\{ \frac{E_z(z)}{\varepsilon_t \vec{E}_z(0)} - \left[\frac{1}{\varepsilon_a}\theta(-z) + \frac{1}{\varepsilon_t}\theta(z)\right] \right\} \tag{5.42a}$$

$$= \frac{\varepsilon_a}{\varepsilon_a - \varepsilon_t} \int\limits_{\xi_1}^{\xi_2} dz\, z\frac{d}{dz}\left(\frac{E_z(z)}{\vec{E}_z(0)}\right) . \tag{5.42b}$$

which follows (for $a = 0$) directly from (5.34) and the definition (5.9b) of $n_z^{(1)}(0)$, since within the LWL $D_z(z) = \vec{D}_z(0) = \varepsilon_t \vec{E}_z(0)$ is constant in the surface region, so that the reference field there reduces to $E_z^0(z;0) = [\varepsilon_a^{-1}\theta(-z) + \varepsilon_t^{-1}\theta(z)]\vec{D}_z(0)$.

According to our general assumptions, for $\omega > \omega_p$ the value of ξ_2 in (5.42) has to be chosen so deep inside the metal that plasma waves excited at the surface are already damped out at $z \approx \xi_2$. In order to have ξ_2 close to the surface ($|\xi_2 - \xi_1|\omega/c$ $\ll 1$) a sufficiently large damping must be assumed. We now discuss briefly the case of weak damping. Then, for $\omega > \omega_p$, $E_z(z)$ contains a weakly damped plasmon contribution $E_z^{pl}(z) = C_{pl}\exp(izp_{pl})$ which extends deep into metal, ξ_2 in (5.42) becomes large, and the replacement of $E_z^0(z;0)/\vec{D}_z(0)$ by the square bracket in (5.42a) is no longer correct, whereas (5.34) with (5.9b), which extends the z integral to infinity, can still be used to define $d_\perp(\omega)$. We recall the basic idea of this chapter: integrate over the rapidly varying surface field in order to derive generalized boundary conditions for the known asymptotic fields. In this spirit we may redefine ξ_2, so that $E_z(z) \approx \vec{E}_z(z) + E_z^{pl}(z)$ for $z \geqslant \xi_2$. Then we can again use (5.42a) with ξ_2 close to the surface, but we have to add

$$\int\limits_{\xi_2}^{\infty} dz\, E_z^{pl}(z)/\varepsilon_t\vec{E}_z(0) = i[C_{pl}/p_{pl}\,\varepsilon_t\,\vec{E}_z(0)]\exp(i\xi_2 p_{pl})$$

to the integral in (5.42a). Formulas of this type have been published by FEIBELMAN /3.84, 86; 5.1/. Thus, in the presence of weakly damped plasmons, reduction of the integral in (5.42) to a narrow surface region is possible, but requires the exact knowledge of the wavenumber p_{pl} of the plasma wave (which is a bulk property /4.30/) and of its amplitude and phase, given by the complex quantity C_{pl} (which contains surface information /4.30/). It has also been suggested /4.14; 5.2/ to include for $\omega > \omega_p$ the plasmon field directly in the reference field (5.1), so that the remaining contributions to $d_\perp(\omega)$ come only from a narrow surface region. Again, this can only work, if the correct strength and phase of the plasmon field is taken into account, which means that in addition to $d_\perp$ and $d_\parallel$ a further complex function of frequency is needed to describe the surface response for $\omega > \omega_p$. To our knowledge a reformulation of the present surface response formalism including plasmon fields explicitly in the set of asymptotic fields has not actually been carried out so far. In the following we will assume that the effects of plasma waves are implicitly included in $d_\perp(\omega)$.

In order to give a simple physical interpretation of $d_\perp(\omega)$, FEIBELMAN /3.84/ emphasized that $\mathrm{Re}\{d_\perp\}$ is the centroid of the induced charge and measures an optical "surface position". Indeed, owing to the rapid variation of $E_z(z)$ in the surface regime, one has in the LWL $|ik_x E_x| \ll |dE_z/dz|$ and $4\pi\rho^{ind} = \nabla\cdot\mathbf{E} \approx dE_z/dz$, so that, for $\omega < \omega_p$,

$$4\pi \int_{-\infty}^{+\infty} dz\ \rho^{ind}(z) = E_z(\xi_2) - E_z(\xi_1) = (1 - \varepsilon_t/\varepsilon_a)E_z^>(0) \tag{5.43}$$

and from (5.42b),

$$d_\perp(\omega) = \int_{-\infty}^{+\infty} dz\ z\,\rho^{ind}(z) \Big/ \int_{-\infty}^{+\infty} dz\ \rho^{ind}(z) \quad . \tag{5.44}$$

If the induced charge density $\rho^{ind}(z)$ is dominated by a single peak, then $\mathrm{Re}\{d_\perp(\omega)\}$ measures the position of this peak and Feibelman's interpretation holds. But it has been pointed out /3.31 - 34 ; 4.30/ that in a diffuse surface a strong electric field can be induced owing to a local plasmon excitation. Then $E_z(z)$ can be strongly peaked in the surface region /3.34 ; 4.30/ so that large positive *and* negative charges are induced, and $\mathrm{Re}\{d_\perp\}$ measures the dipole moment rather than the position of the induced charges.

The imaginary part of $d_\perp(\omega)$ is important for the total power absorption. We refer to FEIBELMAN's review /3.84/ for a detailed discussion.

5.4 Local Model for the Surface Region

Some aspects of nonlocal optics in the LWL resemble the local theory, and the surface functions $d_\perp(\omega)$ and $d_\parallel(\omega)$ can be expressed in terms of a local three layer model with an anisotropic surface layer. We want to emphasize, however, that this is purely formal and a local model is neither helpful for computation nor for understanding of nonlocal surface effects. To make this clear, we evaluate (in this section) the surface integrals for the local model and compare (in Sect. 5.5) the results with those of a nonlocal calculation. First, we show that in general $d_\perp(\omega)$ and $d_\parallel(\omega)$ can be expressed in terms of layer thickness and dielectric constants of a local three layer model with spatially constant values of the dielectric functions within the layers.

In Sect. 5.2 we have exploited the slow variation of $E_x(z)$ and $D_z(z)$, the field components which are continuous in classical optics, to establish the quasi-local relations

$$D_x(z) = \varepsilon_{xx}(z)E_x(z) \quad , \qquad E_z(z) = \varepsilon_{zz}^{-1}(z)D_z(z) \quad . \tag{5.45}$$

Whereas it is straightforward to calculate the integrals $\varepsilon_{\mu\mu}(z) = \int dz' \varepsilon_{\mu\mu}(z,z')$ from a given microscopic model for the dielectric tensor, the computation of $\varepsilon_{zz}^{-1}(z)$ requires the solution of Maxwell's equations. Contrary to a truely local model, in general $\varepsilon_{zz}^{-1}(z) \neq 1/\varepsilon_{zz}(z)$, and (5.45) cannot be used to solve the problem but only to express the solution in a convenient form.

The contributions to the integrals (5.35 , 36) for $d_\perp(\omega)$ and $d_\parallel(\omega)$ come from a surface layer of thickness $\tilde{d}$, which we assume to be located at $\xi - \tilde{d} < z < \xi$. Since outside this layer the response is local, we may assume $\varepsilon_{zz}^{-1}(z) = 1/\varepsilon_a$, $\varepsilon_{zz}(z) = \varepsilon_a$ for $z < \xi - \tilde{d}$ and $\varepsilon_{zz}^{-1}(z) = 1/\varepsilon_t$, $\varepsilon_{zz}(z) = \varepsilon_t$ for $z > \xi$. Introducing the mean value

$$\langle \varepsilon_{zz}^{-1} \rangle = \frac{1}{\tilde{d}} \int\limits_{\xi-\tilde{d}}^{\xi} dz \; \varepsilon_{zz}^{-1}(z) \quad , \qquad \langle \varepsilon_{xx} \rangle = \frac{1}{\tilde{d}} \int\limits_{\xi-\tilde{d}}^{\xi} dz \; \varepsilon_{xx}(z) \quad , \tag{5.46}$$

we may write (5.35 , 36) for $0 \leqslant \xi \leqslant \tilde{d}$ as

$$d_\perp(\omega) = -\frac{\langle \varepsilon_{zz}^{-1} \rangle - 1/\varepsilon_a}{1/\varepsilon_t - 1/\varepsilon_a} \tilde{d} + \xi \tag{5.47}$$

and

$$d_\parallel(\omega) = -\frac{\langle \varepsilon_{xx} \rangle - \varepsilon_a}{\varepsilon_t - \varepsilon_a} \tilde{d} + \xi \quad , \tag{5.48}$$

respectively.

Obviously the same values for $d_\perp(\omega)$ and $d_\parallel(\omega)$ result, if we assume a local three layer model with constant values $\varepsilon_{xx}(z) = \langle \varepsilon_{xx} \rangle$ and $\varepsilon_{zz}(z) = 1/\langle \varepsilon_{zz}^{-1} \rangle$ in the surface layer $\xi - \tilde{d} < z < \xi$. Consequently, in the LWL the optical properties of the surface region - including nonlocal effects - can be described by a local three layer model (dielectric - surface layer - metal), but in general with anisotropic optical constants of the surface layer, since there is no reason to expect that the values of $\langle \varepsilon_{xx} \rangle$ and $1/\langle \varepsilon_{zz}^{-1} \rangle$ are equal.

PLIETH and NAEGELE /5.6/ derived these results on the basis of a truely local model of the form (5.45) with a smoothly varying dielectric function in the surface region, using the Alkemade - Drude derivation /5.3 - 5/ of boundary conditions for the elctromagnetic field. They emphasized the fact that, owing to the necessary anisotropy $\langle \varepsilon_{zz}^{-1} \rangle \neq 1/\langle \varepsilon_{xx} \rangle$ of the equivalent local three layer step-model, it is impossible to determine the thickness d of the surface layer from optical measurements, which, for given frequency, can yield only two complex quantities, e.g., $d_\perp(\omega)$ and $d_\parallel(\omega)$.

Moreover, Plieth and Naegele demonstrated that improper assumptions on the position and width of the surface layer can lead to unreasonable results for the parameters $<\varepsilon_{xx}>$ and $<\varepsilon_{zz}^{-1}>$, and they preferred to present their surface data without referring to a local three layer model.

The anisotropic local three layer model contains more parameters than can be determined by reflection experiments in the LWL. However, it cannot be used beyond that limit if nonlocal effects play a role. To show this explicitly, we now evaluate the surface integrals (5.9) for the three layer model with the local dielectric function ($\xi = d$)

$$
\varepsilon_{\mu\nu}(z) = \delta_{\mu\nu} \cdot \begin{cases} \varepsilon_a & , & z < 0 \\ \varepsilon_{\mu\mu}^S & , & 0 < z < d \\ \varepsilon_t & , & d < z \end{cases} . \tag{5.49}
$$

For the case of p *polarization*, incident and reflected field are given by (5.3) ($z < 0$) and the transmitted wave in the metal ($z > d$) is given by (5.4). In the surface layer, $0 < z < d$, the dispersion relation is

$$
\frac{k_x^2}{\varepsilon_{xx}^S} + \frac{p_S^2}{\varepsilon_{zz}^S} = \frac{\omega^2}{c^2} \quad , \tag{5.50}
$$

and, since $\varepsilon_{xx}^S \neq \varepsilon_{zz}^S$, only $\mathbf{D}^S$ is a transverse field,

$$
D_x^S(z) = D_{x+}e^{izp_S} + D_{x-}e^{-izp_S} \quad , \tag{5.51a}
$$

$$
D_z^S(z) = -\frac{k_x}{p_S}(D_{x+}e^{izp_S} - D_{x-}e^{-izp_S}) \quad , \tag{5.51b}
$$

not $\mathbf{E}^S$. The standard matching conditions, E_x and D_z continuous at $z = a$ and $z = d$, determine the fields completely, and one easily calculates the reflection amplitude r_p.

The surface integrals (5.9) reduce, for $a = 0$, to

$$
\delta_x^{(n)}(0) = \int_0^d dz \, z^{n-1}[D_x^S(z) - \varepsilon_t E_x^t e^{izp_t}]/E_x^t \quad , \tag{5.52}
$$

which yields

$$
\delta_x^{(1)}(0) = \frac{i\varepsilon_t}{p_t}\left\{\left(\cos p_s d - \frac{\varepsilon_{xx}^S p_t}{\varepsilon_t p_s} i \sin p_s d\right)e^{ip_t d} - 1\right\} \quad , \tag{5.53a}
$$

$$\delta_x^{(2)}(0) = \left[\frac{\varepsilon_{xx}^s}{p_s^2}(1 - \cos p_s d) + \frac{\varepsilon_t}{p_s p_t}\, i \sin p_s d\right] e^{ip_t d} - \frac{\varepsilon_t}{p_t^2}\left(e^{ip_t d} - 1\right) , \qquad (5.53b)$$

and to

$$\eta_z^{(1)}(0) = \int_0^d dz \; [E_z^s(z) - \overset{>}{E_z}(z)]/[\varepsilon_t \overset{>}{E_z}(0)] \qquad\qquad\qquad (5.54)$$

$$= \frac{1}{ip_s \varepsilon_{zz}^s}\left[\frac{p_t \varepsilon_{xx}^s}{p_s \varepsilon_t}(1 - \cos p_s d) + i \sin p_s d\right] e^{ip_t d} - \frac{1}{ip_t \varepsilon_t}\left(e^{ip_t d} - 1\right).$$

Inserting (5.53 , 54) into (5.11 , 12), one reobtains the correct value for the re-
flection amplitude r_p. In the LWL, $|p_t d| \ll 1$, $|p_s d| \ll 1$, we expand (5.53 , 54):

$$\delta_x^{(1)}(0) = (\varepsilon_{xx}^s - \varepsilon_t)d + i[p_t \varepsilon_{xx}^s - \tfrac{1}{2}\varepsilon_t(p_t + p_s^2/p_t)]d^2 + \ldots , \qquad (5.55a)$$

$$\delta_x^{(2)}(0) = \tfrac{1}{2}(\varepsilon_{xx}^s - \varepsilon_t)d^2 + \ldots , \qquad\qquad\qquad (5.55b)$$

$$\eta_z^{(1)}(0) = \left(\frac{1}{\varepsilon_{zz}^s} - \frac{1}{\varepsilon_t}\right)d - \frac{ip_t}{\varepsilon_t}[\tfrac{1}{2} - (\varepsilon_t - \tfrac{1}{2}\varepsilon_{xx}^s)/\varepsilon_{zz}^s]d^2 + \ldots . \qquad (5.56)$$

The leading order terms of (5.55a , 56) are in agreement with (5.48 , 47), respectively, for $\xi = \tilde{d} = d$ [cf. (5.28 , 34)]. If the optical constants of surface layer and
bulk metal are equal, $\varepsilon_{xx}^s = \varepsilon_{zz}^s = \varepsilon_t$, then $\delta_x^{(n)}(0) = \eta_z^{(1)}(0) = 0$, as it should be.
   For s *polarization* the field in the dielectric, $z < 0$, and in the metal, $z > d$,
is given by (5.14 , 15), respectively, and in the surface layer $0 < z < d$ we write

$$E_y(z) = E_+ e^{izp_s} + E_- e^{-izp_s} , \qquad\qquad\qquad (5.57)$$

where now $k_x^2 + p_s^2 = \varepsilon_{yy}^s \omega^2/c^2$. The standard matching conditions, E_y and $B_x =$
$(ic/\omega)dE_y/dz$ continuous at $z = 0$ and $z = d$, determine field and reflection amplitude r_s. From (5.9) we obtain

$$\delta_y^{(n)}(0) = \int_0^d dz\; z^{n-1}\left[\varepsilon_{yy}^s(E_+ e^{izp_s} + E_- e^{-izp_s}) - \varepsilon_t E_t e^{izp_t}\right]/E_t , \qquad (5.58)$$

and explicitly

$$\delta_y^{(1)}(0) = \frac{\varepsilon_{yy}^s}{ip_s}\left[\frac{p_t}{p_s}(1 - \cos p_s d) + i \sin p_s d\right] e^{ip_t d} - \frac{\varepsilon_t}{ip_t}(e^{ip_t d} - 1) , \qquad (5.59a)$$

$$\delta_y^{(2)}(0) = \frac{\varepsilon_{yy}^s}{p_s^2}\left(1 - \cos p_s d + \frac{p_t}{p_s} i \sin p_s d - ip_t d\right)e^{ip_t d}$$

$$\hspace{6cm} (5.59b)$$

$$- \frac{\varepsilon_t}{p_t^2}[(1 - ip_t d)e^{ip_t d} - 1] \quad .$$

The corresponding equations for $n_y^{(n)}(0)$ are obtained from (5.58 , 59), if both ε_{yy}^s and ε_t are replaced by $1/\varepsilon_t$. Inserting these results into (5.17 , 19), one obtains the correct value of the reflection coefficient r_s which, of course, can also be calculated without referring to $\delta_y^{(n)}$ and $n_y^{(n)}$.

For $|p_s d| \ll 1$, $|p_t d| \ll 1$, we get the expansions

$$\delta_y^{(1)}(0) = (\varepsilon_{yy}^s - \varepsilon_t)(d + \frac{i}{2} p_t d^2) + \ldots \quad , \hspace{2cm} (5.60a)$$

$$\delta_y^{(2)}(0) = (\varepsilon_{yy}^s - \varepsilon_t)(\frac{1}{2} d + \frac{i}{3} p_t d^3) + \ldots \quad , \hspace{2cm} (5.60b)$$

$$n_y^{(1)}(0) = -\frac{1}{2\varepsilon_t}(p_s^2 - p_t^2)(\frac{1}{3} d^3 + \frac{i}{4} p_t d^4) + \ldots \quad , \hspace{1cm} (5.61a)$$

$$n_y^{(2)}(0) = -\frac{1}{24\varepsilon_t}(p_s^2 - p_t^2)d^4 + \ldots \quad = \frac{1}{24}(1 - \frac{\varepsilon_{yy}^s}{\varepsilon_t}\frac{\omega^2}{c^2})d^4 + \ldots \quad . \hspace{0.5cm} (5.61b)$$

In the LWL only $\delta_y^{(1)}(0) = (\varepsilon_{yy}^s - \varepsilon_t)d$ must be retained in agreement with the general discussion given in Sect. 5.3. If the surface layer has uniaxial anisotropy, $\varepsilon_{xx}^s = \varepsilon_{yy}^s \neq \varepsilon_{zz}^s$, $\delta_y^{(n)}(0) = \delta_x^{(n)}(0)$ holds to leading order, and the reflection amplitudes can be written in the form (5.26 , 39) with

$$d_\perp(\omega) = \frac{(1/\varepsilon_{zz}^s) - (1/\varepsilon_t)}{(1/\varepsilon_a) - (1/\varepsilon_t)}d \quad , \hspace{3cm} (5.62)$$

$$d_\parallel(\omega) = \frac{\varepsilon_{xx}^s - \varepsilon_t}{\varepsilon_a - \varepsilon_t}d \quad , \hspace{3.5cm} (5.63)$$

in agreement with (5.47 , 48) for $\langle\varepsilon_{zz}^{-1}\rangle = 1/\varepsilon_{zz}^s$, $\langle\varepsilon_{xx}\rangle = \varepsilon_{xx}^s$, $\xi = \tilde{d} = d$.

The explicit results of this section demonstrate that the quantities $\delta_\mu^{(n)}$, $n_\mu^{(n)}$, introduced in Sect. 5.2 in order to formulate boundary conditions for the reference fields, are in general complicated functions of the angle of incidence θ_a. Only in the LWL they become simple, independent of θ_a, and two of them are sufficient to describe the reflection amplitudes. In the following section we will use these explicit results in order to show that, beyond the LWL, the reflection from nonlocal systems cannot be described in terms of the parameters of a local three layer model.

5.5 Nonlocal Layer Model in the Hydrodynamic Approximation

In this section we illustrate the formalism of Sect. 5.2,3 for the three layer model including nonlocal effects in two adjacent layers within the hydrodynamic approximation, which has already been used in Chap. 2 without explicit reference to the dielectric tensor $\varepsilon_{\mu\nu}(z,z')$ and its inverse, and we give in the LWL explicit analytical expressions for the surface functions $d_\perp(\omega)$ and $d_\parallel(\omega)$. These results should be useful for the interpretation of optical investigations of metal films deposited on a metallic substrate, where nonlocal effects may be important in both the film and the substrate /3.52/. The explicit result for $d_\perp(\omega)$ will be used for the discussion of "multipole surface plasmons" in Sect. 5.6 and gives an idea about origin and physical meaning of the frequency-dependence of $d_\perp$.

We consider a metallic surface layer ($0 < z < d$) separating a metallic halfspace ($z > d$) from the adjacent dielectric ($z < 0$) with dielectric constant ε_a. Both metallic layers are allowed to sustain longitudinal fields by the choice of model dielectric functions described in Sect. 2.7. The longitudinal dielectric function in the surface layer is written as

$$\varepsilon_{\ell s}(q,\omega) = \varepsilon_{bs}(\omega) - \frac{\omega_{ns}^2}{\omega(\omega + i\gamma_s) - \beta_s q^2} \quad , \tag{5.64a}$$

including spatial dispersion through a Drude-like "free"-electron term in addition to a nondispersive, local term $\varepsilon_{bs}(\omega)$ which takes "bound" electrons into account. The corresponding transverse dielectric function is

$$\varepsilon_{ts} = \varepsilon_{\ell s}(0,\omega) \quad . \tag{5.64b}$$

Material parameters referring to the bulk metal ($z > d$) are denoted by the same symbols, but without the subscript "s". We use the additional boundary conditions introduced in Chap. 2. The case of s polarization, in which longitudinal fields are not excited, reduces to the local treatment of Sect. 5.4 with $\varepsilon_{yy}^s = \varepsilon_{ts}$. In the present section we consider only the case of p polarization.

The electric field in the dielectric ($z < 0$) is given by (5.3), the transverse field in the bulk metal ($z \geqslant d$), $\mathbf{E}^t = \vec{E}$, is given by (5.4). The total field $\mathbf{E} = \mathbf{E}^t + \mathbf{E}^\ell$ in the metallic halfspace contains in addition to $\mathbf{E}^t$ the longitudinal field

$$E_x^\ell(z) = E_x^\ell e^{izp_\ell} \quad , \qquad E_z^\ell(z) = \frac{p_\ell}{k_x} E_x^\ell(z) \quad , \tag{5.65}$$

with $k_x^2 + p_\ell^2 = q_\ell^2$, where $q_\ell(\omega)$ is defined by $\varepsilon_\ell(q_\ell,\omega) = 0$, and both the transverse wavenumber p_t and the longitudinal wavenumber p_ℓ have non-negative real and imaginary parts. In the surface layer ($0 < z < d$) the transverse field is written as

$$E^{ts}_{\begin{Bmatrix}x\\z\end{Bmatrix}}(z) = \begin{Bmatrix}1\\-k_x/p_{ts}\end{Bmatrix} (E^{ts}_{x+}e^{izp_{ts}} \pm E^{ts}_{x-}e^{-izp_{ts}}) \quad , \tag{5.66}$$

with $k_x^2 + p_{ts}^2 = \varepsilon_{ts}\omega^2/c^2$ and upper (lower) symbols referring to the x (z) component. Similar, the longitudinal field in the layer is given by

$$E^{\ell s}_{\begin{Bmatrix}x\\z\end{Bmatrix}}(z) = \begin{Bmatrix}1\\p_{\ell s}/k_x\end{Bmatrix} (E^{\ell s}_{x+}e^{izp_{\ell s}} \pm E^{\ell s}_{x-}e^{-izp_{\ell s}}) \quad , \tag{5.67}$$

with $k_x^2 + p_{\ell s}^2 = q_{\ell s}^2$ and $\varepsilon_{\ell s}(q_{\ell s},\omega) = 0$.

The displacement field is determined by the transverse field only, $\mathbf{D}^S = \varepsilon_{ts}\mathbf{E}^{ts}$ in the surface layer ($0 < z < d$), $\mathbf{D} = \varepsilon_t\mathbf{E}^t$ in the metallic halfspace ($z > d$), and $\mathbf{D} = \varepsilon_a\mathbf{E}^<$ in the dielectric ($z < 0$).

To calculate the coefficients $E^{ts}_{x\pm}$, $E^{\ell s}_{x\pm}$, etc. and the reflection amplitude r_p, we have to exploit the boundary conditions at $z = 0$ and $z = d$. For the calculation of the surface integrals (5.9) the following procedure is convenient.

At the metal-metal interface, $z = d$, we have the two standard boundary conditions "E_x and D_z continuous" and two additional boundary conditions, (2.52a,d). In our notation (5.65 , 67), these ABC are equivalent to require "$\varepsilon_b E_z$ and QE^{ℓ}_x continuous", where E^{ℓ}_x stands for tangential component of the longitudinal electric field and

$$Q = \varepsilon_t/(\varepsilon_b - \varepsilon_t) \tag{5.68}$$

is determined by the transverse and the background dielectric constant on either side of the interface. (If we would replace (5.68) by $Q = q_\ell^2 = k_x^2 + p_\ell^2$, the second ABC would be replaced by the requirement "div $\mathbf{E}$ continuous", which has been suggested as ABC by other authors /3.31 , 69/, whereas $Q = \beta q_\ell^2$ leads to the so called "stress boundary condition" /3.32 , 96/, cf. Sect. 3.11.) Using these four boundary conditions, we can express the coefficients determining the field in the surface layer in terms of the corresponding coefficients in the metallic halfspace. With $\tau = \pm$, we obtain

$$E^{ts}_{x\tau}e^{\tau ip_{ts}d} = \frac{1}{2}\left[\left(1 + \tau\frac{\varepsilon_t p_{ts}}{\varepsilon_{ts}p_t}\right)E^t_x e^{ip_t d} + \left(1 - \frac{Q}{Q_s}\right)E^\ell_x e^{ip_\ell d}\right] \quad , \tag{5.69}$$

$$E^{\ell s}_{x\tau}e^{\tau ip_{\ell s}d} = \frac{1}{2}\left[\tau\frac{k_x^2}{p_t p_{\ell s}}\left(\frac{\varepsilon_t}{\varepsilon_{ts}} - \frac{\varepsilon_b}{\varepsilon_{bs}}\right)E^t_x e^{ip_t d} + \left(\frac{Q}{Q_s} + \tau\frac{\varepsilon_b p_\ell}{\varepsilon_{bs}p_{\ell s}}\right)E^\ell_x e^{ip_\ell d}\right] \cdot \tag{5.70}$$

In order to calculate the surface integrals (5.9), which here reduce to

$$\delta^{(n)}_x(0) = \int_0^d dz\, z^{n-1}[\varepsilon_{ts}E^{ts}_x(z) - \varepsilon_t E^t_x e^{izp_t}]/E^t_x \tag{5.71}$$

108

and

$$\eta_z^{(1)}(0) = \left\{ \int_0^d dz \ [E_z^{ts}(z) + E_z^{\ell s}(z) - \vec{E}_z(z)] + \int_d^\infty dz \ E_z^\ell(z) \right\} / [\varepsilon_t \vec{E}_z(0)] \tag{5.72}$$

with $\vec{E}_z(z) = (-k_x P_t) E_x^t \exp(ip_t z)$, we need in addition to (5.69, 70) the coefficient E_x^ℓ in terms of E_x^t. This relation is provided by the ABC at the dielectric-metal interface ($z = 0$): the condition that the normal component of the free electron current must vanish there yields

$$\lambda \equiv (E_x^\ell / E_x^t) \exp[i(p_\ell - p_t)d]$$

$$= \frac{k_x^2}{p_t p_\ell} \frac{\left(1 - \frac{\varepsilon_{ts}}{\varepsilon_{bs}}\right)\left[\frac{\varepsilon_t}{\varepsilon_{ts}} \cos p_{ts}d - \frac{p_t}{p_{ts}} i \sin p_{ts}d\right] - \left(\frac{\varepsilon_t}{\varepsilon_{ts}} - \frac{\varepsilon_b}{\varepsilon_{bs}}\right) \cos p_{\ell s}d}{\frac{\varepsilon_b}{\varepsilon_{bs}} \cos p_{\ell s}d - \frac{p_{\ell s}Q}{p_\ell Q_s} i \sin p_{\ell s}d + \frac{k_x^2}{p_\ell p_{ts}}\left(1 - \frac{\varepsilon_{ts}}{\varepsilon_{bs}}\right)\left(1 - \frac{Q}{Q_s}\right) i \sin p_{ts}d} . \tag{5.73}$$

Of course one can use the standard boundary conditions at $z = 0$ to solve directly for the reflection amplitude r_p, which can be obtained in the form

$$r_p = (A - B)/(A + B) \quad , \tag{5.74a}$$

$$A = \varepsilon_t p_a \left\{ \cos p_{ts}d - \frac{\varepsilon_{ts}p_t}{\varepsilon_t p_{ts}}[1 + (1 - \frac{Q}{Q_s})\lambda] i \sin p_{ts}d \right\} \quad , \tag{5.74b}$$

$$B = \varepsilon_a p_t \left\{ [1 + (1 - \frac{Q}{Q_s})\lambda] \cos p_{ts}d - \frac{\varepsilon_t p_{ts}}{\varepsilon_{ts}p_t} i \sin p_{ts}d \right.$$

$$\left. - \frac{k_x^2}{p_t p_{\ell s}}\left(\frac{\varepsilon_t}{\varepsilon_{ts}} - \frac{\varepsilon_b}{\varepsilon_{bs}}\right) i \sin p_{\ell s}d + \lambda\left(\frac{Q}{Q_s} \cos p_{\ell s}d - \frac{\varepsilon_b p_\ell}{\varepsilon_{bs}p_{\ell s}} i \sin p_{\ell s}d\right) \right\} \tag{5.74c}$$

Apart from notation, this agrees with (3.14 - 30), $r_p = R_p$.

Here, however, we want to illustrate the formalism of Sect. 5.2. For the evaluation of the surface integrals (5.71, 72), the standard boundary conditions are not needed. We obtain

$$\delta_x^{(1)}(0) = i \frac{\varepsilon_t}{p_t}\left\{\left[\cos p_{ts}d - \frac{\varepsilon_{ts}p_t}{\varepsilon_t p_{ts}}\left(1 + (1 - \frac{Q}{Q_s})\lambda\right) i \sin p_{ts}d\right]e^{ip_t d} - 1\right\} , \tag{5.75a}$$

$$\delta_x^{(2)}(0) = \frac{\varepsilon_{ts}}{p_{ts}^2}\left\{[1 + (1 - \frac{Q}{Q_s})\lambda](1 - \cos p_{ts}d) + \frac{\varepsilon_t p_{ts}}{\varepsilon_{ts}p_t} i \sin p_{ts}d\right\}e^{ip_t d}$$

$$- \frac{\varepsilon_t}{p_t^2}(e^{ip_t d} - 1) \quad , \tag{5.75b}$$

and

$$\eta_z^{(1)}(0) = \frac{e^{iP_td}}{\varepsilon_t}\left\{\frac{P_t}{iP_{ts}^2}\left[1 - \cos P_{ts}d + \frac{\varepsilon_t P_{ts}}{\varepsilon_{ts}P_t} i \sin P_{ts}d\right] - \left(\frac{\varepsilon_t}{\varepsilon_{ts}} - \frac{\varepsilon_b}{\varepsilon_{bs}}\right)\frac{\sin P_{\ell s}d}{P_{\ell s}}\right.$$

$$- \frac{1}{iP_t}(1 - e^{-iP_td}) + \lambda\frac{P_t}{ik_x^2}\left[\frac{k_x^2}{P_{ts}^2}\left(1 - \frac{Q}{Q_s}\right)(1 - \cos P_{ts}d)\right. \tag{5.76}$$

$$\left.\left. - \frac{Q}{Q_s}(1 - \cos P_{\ell s}d) + 1 - \frac{\varepsilon_b P_\ell}{\varepsilon_{bs}P_{\ell s}} i \sin P_{\ell s}d\right]\right\} \quad .$$

These results are exact and, if inserted into (5.11, 12) for a = 0, they lead back
to the expression (5.74) for r_p. It should be noted that in this last step the generalized boundary conditions (5.10) are implied which, similar to the standard boundary conditions, were derived from Maxwell's equations. For thick metallic layers on
metal substrates /3.35/ it is necessary to work with the exact (within HD) formula
(5.74) (see also the discussion of surface plasmons in Sect. 3.4), but for thin
surface layers and clean surfaces simplifications are possible. We now consider
some special and limiting cases.

5.5.1 No Surface Layer

If the optical constants of the surface layer are the same as those of the bulk
metal, $\varepsilon_{ts} = \varepsilon_t$, $\varepsilon_{bs} = \varepsilon_b$, $P_{\ell s} = P_\ell$, etc., the interface at z = d has no physical
effect. Then (5.69, 70) yield $E_{x-}^{ts} = E_{x-}^{\ell s} = 0$, $E_{x+}^{ts} = E_x^t$, $E_{x+}^{\ell s} = E_x^\ell$, and (5.73) reduces
to

$$\lambda = \frac{k_x^2}{P_tP_\ell}\left(1 - \frac{\varepsilon_t}{\varepsilon_b}\right)e^{i(P_\ell - P_t)d} \quad , \tag{5.77}$$

so that (4.75, 76) yield

$$\delta_x^{(1)}(0) = \delta_x^{(2)}(0) = 0 \quad , \tag{5.78}$$

$$\eta_z^{(1)}(0) = \frac{i}{P_\ell}\left(\frac{1}{\varepsilon_b} - \frac{1}{\varepsilon_t}\right) \quad , \tag{5.79}$$

whereas (5.74) simplifies to

$$r_p = \frac{\varepsilon_tP_a - \varepsilon_aP_t - \varepsilon_ak_x^2(1 - \varepsilon_t/\varepsilon_b)/P_\ell}{\varepsilon_tP_a + \varepsilon_aP_t + \varepsilon_ak_x^2(1 - \varepsilon_t/\varepsilon_b)/P_\ell} \quad , \tag{5.80}$$

which agrees with a result of ABELES and LOPEZ-RIOS /5.9/ and reduces for $\varepsilon_a = \varepsilon_b$
= 1 to the free-electron formula (3.6). According to (5.28, 34), $\delta_y^{(1)}(0) = \delta_x^{(1)}(0)$

and (5.79) yields

$$d_{\parallel}(\omega) = 0 \quad , \qquad d_{\perp}(\omega) = \frac{i}{p_{\ell}} \frac{1/\varepsilon_b - 1/\varepsilon_t}{1/\varepsilon_a - 1/\varepsilon_t} \quad . \tag{5.81}$$

Inserting this into the LWL-formula (5.37), we reobtain (5.80). This shows that for the present nonlocal model (HD) of a homogeneous metallic halfspace (5.37,81) are valid beyond the LWL.

It is interesting to compare (5.81) with (5.62,63) for the local three layer model, and to find out how we have to choose the parameters of that local model in order to reproduce the reflection properties of the present nonlocal model. Obviously, we have to set $\varepsilon_{xx}^S = \varepsilon_t$. A simple choice is $d = i/p_{\ell}$ and $\varepsilon_{zz}^S = \varepsilon_b$. For small damping and frequencies below the plasma frequency, this choice seems reasonable, since p_{ℓ} is imaginary $[0 \leqslant \mathrm{Re}\{p_{\ell}\} \ll \mathrm{Im}\{p_{\ell}\}$ if $\mathrm{Im}\{\varepsilon_t\} \ll -\mathrm{Re}\{\varepsilon_t\}]$ and i/p_{ℓ} is the decay length of the longitudinal field at the surface, i.e., of the plasma wave. Then the thickness of the effective optically anisotropic surface layer, which shall simulate nonlocal effects by a local model, is just the penetration depth of the longitudinal field, so that deviations from local metal optics occur only in this surface layer.

But this $d = i/p_{\ell}$ depends on frequency and becomes large at the plasma frequency. Then the LWL formula (5.62) is no longer valid, and we should compare (5.56) with (5.79). Owing to the first order identification $d = i/p_{\ell}$, $\varepsilon_{zz}^S = \varepsilon_b$, $\varepsilon_{xx}^S = \varepsilon_t$ the d^2 term of (5.56) should vanish. This requires $\varepsilon_{xx}^S = 2\varepsilon_t - \varepsilon_b$ or $\varepsilon_t = \varepsilon_b$. Thus, we see that *beyond the LWL the optical properties* of the present simple nonlocal model *cannot consistently be described by a local three layer model*.

Also within the long wavelength limit the choice $d = i/p_{\ell}$ is not really satisfactory. For finite damping this is always a complex quantity and above the plasma frequency its imaginary part dominates its real part. On the other hand, if we choose a constant, frequency-independent value of d, comparison of (5.81) and (5.62) yields a value of ε_{zz}^S which depends strongly on this value of d and also on the frequency. Obviously such a choice of parameters is not helpful for an understanding of the optical response of the nonlocal system.

We conclude that, although it is in the LWL formally possible to simulate reflection properties of a nonlocal system by a local three layer model, there is no physical reason to do so, since the parameters ε_{xx}^S, ε_{zz}^S and d of the local model are not uniquely defined and have no clear physical meaning.

5.5.2 Local Limits

The general formulas (5.69 - 76) apply also to the special cases that the dielectric response of the surface layer and/or the metallic halfspace is local, but it requires some care to pass from the nonlocal case to the local limit.

If we want to neglect nonlocal effects in the surface layer, we have to omit spatial dispersion, i.e., to set $\beta_s = 0$ in (5.64a), and, thereby, to omit longitudinal fields in the layer. But if we naively perform the limit $\beta_s \to 0$, which implies $q_{\ell s} \to \infty$, $p_{\ell s} \to \infty$, we get wrong results which are different from those of a calculation which neglects longitudinal fields in the layer from the beginning. Similarly, if we want to retain nonlocality in the surface layer but not in the metallic halfspace, we get a wrong answer if we take the limit $\beta \to 0$, $p_\ell \to \infty$. The difficulty is related to the second ABC (2.52d) ($Q E_x^\ell$ continuous) at the metal-metal interface, which couples the induced charges on both sides of the interface. If there are longitudinal fields only on one side of the interface, this ABC cannot be satisfied. But in the limit $\beta \to 0$, which leaves the value $Q = \varepsilon_t/(\varepsilon_b - \varepsilon_t)$ unchanged, this ABC is not abandoned.

Fortunately there is a simple alternative way to switch off the spatial dispersion in (5.64a). We may, for $0 < x < 1$, replace ω_{ns}^2 in (5.64a) by $x\omega_{ns}^2$ and include at the same time an additive contribution $-(1 - x)\omega_{ns}^2/[\omega(\omega + i\gamma_s)]$ in the nondispersive part $\varepsilon_{bs}(\omega)$. This procedure leaves the transverse dielectric constant ε_{ts} unchanged and leads in the limit $x \to 0$ to the local case.

With the redefined ε_{bs}, the correct prescription for taking the local limit of the surface layer ($x \to 0$), is to set $\varepsilon_{bs} = \varepsilon_{ts}$ and $Q_s = \infty$. Then (5.73) reduces to

$$\lambda = \frac{k_x^2}{p_t p_\ell}\left(1 - \frac{\varepsilon_t}{\varepsilon_b}\right) \quad , \tag{5.82}$$

and (5.70) yields $E_{x+}^{\ell s} = E_{x-}^{\ell s} = 0$, the absence of longitudinal fields in the surface layer. Furthermore, (5.76) reduces to

$$\eta_z^{(1)}(0) = \frac{1}{ip_{ts}\varepsilon_{ts}}\left[\frac{\varepsilon_{ts}p_t}{\varepsilon_t p_{ts}}(1 - \cos p_{ts}d) + i \sin p_{ts}d\right]e^{ip_t d}$$

$$- \frac{1}{ip_t\varepsilon_t}(e^{ip_t d} - 1) + \frac{1}{ip_\ell\varepsilon_t}\left(1 - \frac{\varepsilon_t}{\varepsilon_b}\right)\left[1 + \frac{k_x^2}{p_{ts}^2}(1 - \cos p_{ts}d)\right]e^{ip_t d} \quad , \tag{5.83}$$

independent of $p_{\ell s}$.

If we set in addition $\varepsilon_b = \varepsilon_t$, we neglect nonlocal effects in the metallic halfspace, $\lambda = 0$, and $\delta_x^{(n)}(0)$, $\eta_z^{(1)}(0)$ reduce to the local values (5.53, 54) with an isotropic surface dielectric tensor $\varepsilon_{xx}^s = \varepsilon_{zz}^s = \varepsilon_{ts}$. (Note that the last step, $\varepsilon_b \to \varepsilon_t$, has the same effect as taking the limit $p_\ell \to \infty$, or $\beta \to 0$. But this is only

112

true, since we assumed already a local surface-layer response.)

If we want to retain nonlocality in the surface layer but not in the metallic halfspace, we take in (5.69 - 76) the limit $\varepsilon_b \to \varepsilon_t$, $Q \to \infty$. Then E_x^ℓ vanishes according to (5.73), but $\lambda Q/Q_s$ remains finite and becomes independent of Q_s, the quantity which occurs only through the second ABC. It is straightforward to evaluate this limit, but the resulting formulas remain rather lengthy. We give explicit results only for the important LWL.

5.5.3 Long Wavelength Limit

We now assume that the thickness d of the surface layer and the wavelength and/or decay length of the plasma waves in both the surface layer and the bulk metal are much smaller than the wavelength and/or decay length of transverse electric fields in any of the space regions, so that $p_{ts}d$, $p_t d$, $k_x d$, $k_x/p_{\ell s}$, k_x/p_ℓ, etc., are much smaller than unity, and $p_{\ell s}$ and p_ℓ are effectively independent of the angle of incidence θ_a [$k_x = (\omega/c)\varepsilon_a^{1/2}\sin\theta_a$]. For free-electron metals, but also for noble metals, this is a good approximation for all frequencies of interest.

In this long wavelength limit (5.73) reduces to

$$\lambda = \frac{k_x^2}{p_t} \cdot \frac{\varepsilon_t[\mu_s + (\mu - \mu_s)\cos p_{\ell s}d]}{p_\ell \varepsilon_b \cos p_{\ell s}d - p_{\ell s}\varepsilon_{bs}(Q/Q_s)\, i\sin p_{\ell s}d} \quad , \tag{5.84}$$

where

$$\mu_s = (\varepsilon_{bs} - \varepsilon_{ts})/\varepsilon_{ts} \quad , \tag{5.85}$$

and similar μ, has been introduced for brevity. According to (5.68), $Q = 1/\mu$ for our ABC. But we retain Q explicitly in order to see the possible effect of other ABC on the final result (e.g. $Q = q_\ell^2 /3.31$, 69/ or $Q = \beta q_\ell^2 /3.32$, 96/).

To leading order in $p_{ts}d$ the nonlocal correction to $\delta_x^{(n)}(0)$ in (5.75) is negligible

$$\delta_x^{(1)}(0) = (\varepsilon_{ts} - \varepsilon_t)d \quad , \qquad \delta_x^{(2)}(0) = 0 \quad , \tag{5.86}$$

and (5.76) reduces after some algebra to

$$\eta_z^{(1)}(0) = \left(\frac{1}{\varepsilon_{ts}} - \frac{1}{\varepsilon_t}\right)d + i\left[\frac{[(2\mu_s - \mu)\dfrac{Q}{Q_s} - \mu_s](1 - \cos p_{\ell s}d)}{p_\ell \varepsilon_b \cos p_{\ell s}d - p_{\ell s}\varepsilon_{bs}(Q/Q_s)i\sin p_{\ell s}d}\right.$$

$$\left. + \frac{-\mu\cos p_{\ell s}d + \mu_s \dfrac{p_\ell \varepsilon_b}{p_{\ell s}\varepsilon_{bs}}i\sin p_{\ell s}d}{p_\ell \varepsilon_b \cos p_{\ell s}d - p_{\ell s}\varepsilon_{bs}(Q/Q_s)i\sin p_{\ell s}d}\right] \quad , \tag{5.87}$$

113

which includes nonlocal effects in both the surface layer and the metallic half-space, and reduces to the local result for $\mu = \mu_s = 0$.

If we set $\varepsilon_{ts} = \varepsilon_t$, $\varepsilon_{bs} = \varepsilon_b$, and $Q_s = Q$, etc., we recover (5.79), the result for a homogeneous metallic halfspace in $z > 0$.

For local response of the surface layer, $\mu_s \rightarrow 0$, $Q_s \rightarrow \infty$, (5.87) reduces to

$$\eta_z^{(1)}(0) = \left(\frac{1}{\varepsilon_{ts}} - \frac{1}{\varepsilon_t}\right)d + \frac{i}{p_\ell}\left(\frac{1}{\varepsilon_b} - \frac{1}{\varepsilon_t}\right) \quad , \tag{5.88}$$

the LWL of (5.84).

If we retain nonlocality only in the surface layer but not in the bulk, $\mu \rightarrow 0$, $Q \rightarrow \infty$, we obtain from (5.87)

$$\eta_z^{(1)}(0) = \left(\frac{1}{\varepsilon_{ts}} - \frac{1}{\varepsilon_t}\right)d - \frac{2}{p_{\ell s}}\left(\frac{1}{\varepsilon_{ts}} - \frac{1}{\varepsilon_{bs}}\right) \tan(\tfrac{1}{2}p_{\ell s}d) \quad , \tag{5.89}$$

which leads with (5.28, 34, 86) to

$$d_\perp - d_\parallel = d\frac{\varepsilon_a}{\varepsilon_t - \varepsilon_a}\left[(\varepsilon_t - \varepsilon_{ts})\left(\frac{1}{\varepsilon_{ts}} - \frac{1}{\varepsilon_a}\right) - \varepsilon_t\left(\frac{1}{\varepsilon_{ts}} - \frac{1}{\varepsilon_{bs}}\right)\frac{\tan\left(\frac{1}{2}p_{\ell s}d\right)}{\frac{1}{2}p_{\ell s}d}\right] . \tag{5.90}$$

Inserting this into the ellipsometry formula (5.40) we obtain a result which has first been published and discussed by ABELES and LOPEZ /5.9/. From (5.89, 90) it becomes obvious that nonlocal effects in the surface layer are most important near its plasma frequency ω_{ps}. For $\omega = \omega_{ps}$, the value $\eta_z^{(1)}(0) = (\varepsilon_{ts}^{-1} - \varepsilon_t^{-1})d$ of the local approximation becomes large and $p_{\ell s}$ becomes small (in the absence of damping: $\varepsilon_{ts} = 0$, $p_{\ell s} = 0$ for $\omega = \omega_{ps}$). Then $\tan(p_{\ell s}d/2)$ reduces to $p_{\ell s}d/2$ and (5.89) yields $\eta_z^{(1)}(0) = (\varepsilon_{bs}^{-1} - \varepsilon_t^{-1})d$, i.e. the nonlocal effects remove the large structure in the ω dependence of $\eta_z^{(1)}(0)$ predicted by the local approximation. The same mechanism is responsible for the large discrepancy between the local and the nonlocal calculation of the reflection coefficient near the plasma frequency of the surface layer shown in Fig. 3.13 of Sect. 3.5. Well below ω_{ps} the nonlocal effects in (5.90) yield only a small correction to the local results, since $p_{\ell s} \approx i|p_{\ell s}|$ and $|p_{\ell s}|d \gg 1$ for typical metals, except for very thin surface layers (of a few Angstroms). For $\omega > \omega_{ps}$ and small damping, $p_{\ell s}$ is real and (5.89, 90) predict strong plasma wave effects for such frequencies for which $p_{\ell s}d/2$ becomes an odd multiple of $\pi/2$. In this case d is an odd multiple of the half plasma wavelength $\lambda_{\ell s} = 2\pi/p_{\ell s}$ of the layer, and standing plasma waves can be excited in the layer. These are the plasma resonances of a thin metallic film predicted by MELNYK and HARRISON /3.5/ and observed by LINDAU and NILSSON /3.6/, as discussed in Sect. 3.2.

The important differences between the nonlocal formula (5.87) and the local approximation for the interpretation of ellipsometry experiments near the plasma frequency of the substrate have recently been emphasized in a numerical study by KEMPA and GERHARDTS /3.34/.

114

In Sect. 3.4 we discussed the interesting question whether standing plasma waves can also be optically excited in a thin metal film on a metallic substrate. Within the present context the answer is given by (5.87). Plasma resonances in the surface layer will occur, if the denominator becomes small, i.e., if

$$\tan(p_{\ell s}d) \approx p_{\ell}\varepsilon_b Q_s/(ip_{\ell s}\varepsilon_{bs}Q) \quad . \tag{5.91}$$

If we neglect damping and assume the bulk plasma frequency ω_p larger than that of the surface layer, ω_{ps}, then, for $\omega_{ps} < \omega < \omega_p$, $p_{\ell s}$ is real whereas $p_{\ell} = i|p_{\ell}|$ is purely imaginary, so that the right hand side of (5.91) is a smoothly varying, real function of ω. Thus, if the values of $p_{\ell s}d$ varies in the interval $\omega_{ps} < \omega < \omega_p$ by more than π, (5.91) is satisfied in this interval at least once, whereas for $\omega < \omega_{ps}$ and $\omega > \omega_p$ (5.91) cannot be satisfied (if we assume both ε_b and ε_{bs} positive). If damping effects are not too large, one expects therefore in the frequency interval between the plasma frequency of the surface layer and that of the bulk metal resonances due to excitation of standing plasma waves in the layer, provided the layer is thick enough and the differenc between the plasma frequencies, $\omega_p - \omega_{ps} > 0$, is large enough. This qualitative discussion applies to the FORSTMANN - STENSCHKE ABC (2.52) $[Q = \varepsilon_t/(\varepsilon_b - \varepsilon_t)]$ and to the BOARDMAN - RUPPIN ABC (3.75 , 76) $(Q = q_{\ell}^2)$ as well. Characteristic differences become apparent, if one assumes the bulk plasma frequency ω_p to be much larger than that of the surface layer, ω_{ps}. Then for $\omega_{ps} < \omega \ll \omega_p$ one finds $p_{\ell s} \ll |p_{\ell}|$ and (5.91) is satisfied for $\cot(p_{\ell s}d) \approx 0$, i.e., if $p_{\ell s}d$ is close to an odd multiple of $\pi/2$, which means that the surface layer contains an odd multiple of one quarter of a plasma wavelength. If the ABC is changed ($Q = q_{\ell}^2$ or $Q = \beta q_{\ell}^2$), (5.91) is satisfied for $\tan(p_{\ell s}d) \approx 0$, i.e. $p_{\ell s}d$ is close to an integer multiple of π, which means that the surface layer accomodates an integer (non-zero) multiple of a half plasma wavelength /5.11/. The results agree with those obtained in Sect. 3.4. This demonstrates that the frequency-dependence of $n_z^{(1)}(0)$ or, equivalently, of $d_\perp$ contains the information about excitation modes in the surface layer.

If we interpret the surface layer as the selvedge region of the metal halfspace, differing from the bulk metal only by a reduced density of free electrons, we deal with the hydrodynamic model used in Sect. 3.9 to discuss surface electromagnetic fields. The "local plasmon" excitation discussed in that context is physically the same thing as a standing plasma wave in the surface layer. Surface resonances of this type are closely related to the "multipole surface plasmons" /3.25 - 27 , 31 ; 5.11/ to be discussed in the following section.

In the preceding discussion, excitation of standing plasma waves in the surface layer was related to pole type singularities of $d_\perp(\omega) - d_\parallel(\omega)$ or, equivalently, of $n_z^{(1)}(0)$ given by (5.87). It must be understood that a genuine singularity of $n_z^{(1)}(0)$

as a function of ω can occur only in the LWL. Such a singularity results from vanishing denominator of the LWL of λ, given by (5.84). From the general definition of λ, (5.73), we understand the meaning of this singularity. The ratio E_x^ℓ/E_x^t of the x component of longitudinal and transverse electric field, being usually of the order of $k_x^2/p_t p_\ell \ll 1$, is now of order unity, since a strong longitudinal field is accompanied with the standing wave. As a consequence, the value of $n_z^{(1)}(0)$ given by (5.76) is, at the resonance frequency, enhanced by a factor of $p_t p_\ell/k_x^2$. In the LWL this enhancement appears as a singularity.

5.6 Surface Plasmons

In this section we generalize the treatment of surface plasmons given in Sect. 3.3 in two directions. First, we express the plasmon dispersion relation in terms of the surface response functions $d_\perp(\omega)$ and $d_\parallel(\omega)$ and obtain a general result which applies not only to the hydrodynamic model but also to microscopic surface models. In this part we follow FEIBELMAN's discussion /3.84/. Second, we extend the discussion to the so called "multipole surface plasmons" (MSP) /3.25 - 27 , 31 , 32/, which are additional surface modes including standing plasma waves in a surface region of low electron density. The possible existence of such modes at clean metal surfaces has been a matter of controversy for more than a decade /3.23 , 25-27 , 32 , 96 ; 5.11 - 13/. Our discussion, following recent work of KEMPA and GERHARDTS /3.34/, will establish a relation between the MSP and the photoemission experiment on aluminum first pointed out by SCHWARTZ and SCHAICH /3.32/ (cf. Sect. 3.4 , 10). The explicit results of Sect. 5.5 will be helpful for a qualitative understanding of the frequency dependence of $d_\perp$. The LWL is taken throughout the whole section.

5.6.1 Feibelman's Treatment

As mentioned in Sect. 3.3, a surface plasmon at a metal/vacuum interface can be defined as an eigensolution of Maxwell's equations with an electric field propagating along the interface and decaying exponentially into both the metal and the vacuum. Such an eigenmode exists if the reflection amplitude for p polarized light becomes (with the notation of Sect. 5.2) singular for

$$P_a = iP_a \quad , \qquad P_t = iP_t \quad , \tag{5.92}$$

and $P_a > 0$, $P_t > 0$. Then finite reflected and transmitted fields, both decaying away from the surface, are possible for vanishing incident field. Inserting (5.92) into (5.37), we obtain as condition for the existence of a surface plasmon that the denominator of

$$r_p = \frac{\varepsilon_t P_a - \varepsilon_a P_t + (\varepsilon_a - \varepsilon_t)(P_a P_t d_\parallel + k_x^2 d_\perp)}{\varepsilon_t P_a + \varepsilon_a P_t + (\varepsilon_a - \varepsilon_t)(P_a P_t d_\parallel - k_x^2 d_\perp)} \tag{5.93}$$

must vanish. From the dispersion of transverse waves,

$$k_x^2 = P_a^2 + \varepsilon_a \omega^2/c^2 = P_t^2 + \varepsilon_t \omega^2/c^2 \quad , \tag{5.94}$$

one has

$$(\varepsilon_t P_a + \varepsilon_a P_t)(P_a - P_t) = (\varepsilon_a - \varepsilon_t)(P_a P_t - k_x^2) \quad . \tag{5.95}$$

If one assumes $|P_a d_\parallel| \sim |k_x d| \ll 1$, then (5.95) shows that for vanishing denominator of (5.93) the difference $P_a P_t - k_x^2$ is much smaller than both $P_a P_t$ and k_x^2 /3.84/. Hence, in the LWL, the condition for a surface plasmon can be written as

$$\varepsilon_t P_a + \varepsilon_a P_t + (\varepsilon_t - \varepsilon_a)(d_\perp - d_\parallel) P_a P_t = 0 \quad . \tag{5.96}$$

This form of the surface plasmon dispersion relation (SPDR) has been derived by FEIBELMAN /3.84/ and in similar form (but with less transparent methods) also by other authors /5.7 , 10/. Since (5.96) depends only on the difference $d_\perp - d_\parallel$, the SPDR is not changed if the surface position is shifted by an amount a, whereas the values of both $d_\perp$ and $d_\parallel$ are changed by that amount a, as is easily seen from (5.35 , 36). This consistency requirement is also satisfied by the form

$$\varepsilon_t P_a + \varepsilon_a P_t + (\varepsilon_t - \varepsilon_a)(d_\perp - d_\parallel) k_x^2 = 0 \quad , \tag{5.97}$$

which is completely equivalent to (5.96). This form (5.97) reduces for the HD with a single step electron density profile ($d_\parallel = 0$, $d_\perp = i/p_\ell$ from (5.81) with $\varepsilon_a = \varepsilon_b$ = 1) exactly to the SPDR (3.24) (with notation $\lambda_0 = iP_a$, $\lambda = iP_t$, $\varepsilon = \varepsilon_t$, $\eta = p_\ell$) which is not restricted to the long wavelength limit.

In the non-retarded ($c \to \infty$) limit, (5.94) yields $P_t = P_a = k_x$, and both (5.96 , 97) reduce to

$$\varepsilon_t + \varepsilon_a + (\varepsilon_t - \varepsilon_a)(d_\perp - d_\parallel) k_x = 0 \quad . \tag{5.98}$$

With the free-electron dielectric constant, $\varepsilon_t = 1 - \omega_p^2/\omega^2$, this takes, for small k_x, the form /3.84/

$$\omega_s = \frac{\omega_p}{\sqrt{\varepsilon_a + 1}}\left[1 + \frac{\varepsilon_a}{\varepsilon_a + 1}(d_\perp - d_\parallel) k_x\right] \quad , \tag{5.99}$$

for the SPDR, where $d_\perp - d_\parallel$ is taken at $\omega = \omega_s \approx \omega_p/(\varepsilon_a + 1)^{1/2}$. For this surface plasmon the induced charge density $\rho^{ind}(z)$ is dominated by a single peak in the

surface region, since $4\pi\rho^{ind}(z) = \nabla \cdot \mathbf{E} \approx dE_z/dz$, and $E_z(z)$ interpolates smoothly between the nearly constant (in the LWL) classical values outside and inside the metal.
Then, according to (5.44), $d_\perp(\omega)$ measures the position of this peak /3.84/, i.e.
the mean position of the induced charge. Sign and value of $d_\perp - d_\parallel$ depend crucially
on the diffuseness of the surface. If the electron density of the unperturbed metal
is assumed to drop at the surface abruptly from the constant bulk value to zero, as
in the simple hydrodynamic model or in the "semi classical infinite barrier" model,
but also if the model density profile is too steep, as for the microscopic infinite
barrier model, the induced charge lies on the metal side of the surface (of the jellium edge for the IBM) and $d_\perp - d_\parallel$ comes out positive. For the soft, self-consistent
Lang-Kohn profile, on the other hand, the induced charge is located essentially
outside the jellium edge and $d_\perp - d_\parallel$ is negative. Within the hydrodynamic model this
can be simulated by a suitably chosen surface layer of reduced electron density. If
retardation effects are included, the negative values of $d_\perp - d_\parallel$ obtained for diffuse
surfaces lead to a plateau in the surface-plasmon dispersion relation as shown in
Fig. 3.6 of Sect. 3.3.

5.6.2 Additional Surface Plasmon Modes

In order to understand how additional surface plasmons can be discussed in terms
of the surface response function $d_\perp(\omega) - d_\parallel(\omega)$, it is instructive to assume that
$d_\perp(\omega)$ exhibits at a certain "resonance frequency" ω_r a pole singularity, e.g.,

$$d_\perp - d_\parallel \approx \frac{a_r \omega_r}{\omega - \omega_r} \qquad \text{for } \omega \approx \omega_r \quad . \tag{5.100}$$

We have discussed in the preceding Sect. 5.5 that this can happen, for instance,
in the nonlocal three layer model if the frequency is larger than the plasma frequency of the surface layer, but less than the plasma frequency of the bulk metal.
Then, in addition to the "regular" surface plasmon, (5.99), the dispersion relation
(5.98) yields a branch with frequency near ω_r (for small positive values of k_x),

$$\omega = \omega_r \left[1 + \frac{\varepsilon_a - \varepsilon_t}{\varepsilon_a + \varepsilon_t} a_r k_x \right] \quad , \tag{5.101}$$

where the bulk dielectric function $\varepsilon_t = 1 - (\omega_p/\omega)^2$ is taken at $\omega = \omega_r$. These additional surface plasmon modes are accompanied with standing plasma waves in the sur-
face region of reduced electron density, as we discussed in the preceding Sect. 5.5.
Several workers /3.25-27, 31, 96; 5.11/ have studied these modes within the hydrodynamic approximation for the electronic response, assuming either a stepped or a
smooth variation of electron density at the surface. BENNETT /3.23/ first pointed
out their existence, and EGUILUZ et al. /3.25-27; 5.11/, who discussed their ap-

pearance for arbitrarily shaped density profile, addressed them as "higher multi-pole" modes, since in the nonretarded limit the total induced charge of these additional modes was found to vanish. If retardation effects are properly taken into account, this is no longer true. BOARDMAN et al. /3.31/ investigated in detail the electric field and the fluctuation charge density in this case and obtained an oscillatory behaviour of the latter, although not a clear multipole structure.

Quantum mechanical RPA calculations of MSP modes in the nonretarded limit have been presented by INGLESFIELD and WIKBORG /5.12/, who used a double step function to simulate the effective potential (not the density) of conduction electrons at an aluminum surface covered with an overlayer of alkali atoms. "Multipole" modes were found for overlayers with a sufficiently extended low density region, but not for a single step potential, which was considered as a reasonable model of uncoated Al.

Having related the existence of a MSP mode to a pole singularity of $d_\perp(\omega)$, we should understand how such a singularity can be consistent with (5.44), telling that $d_\perp(\omega)$ is the "center of gravity" of the fluctuation charge density. Since $d_\perp(\omega)$ is independent of the wavenumber k_x, we can calculate $d_\perp(\omega)$ from the fields excited by an external plane wave impinging on the surface rather than by the fields related to a surface eigenmode with $k_x > \omega/c$. According to (5.43), the total fluctuation charge, determined by the transverse fields far from the surface, is insensitive to details of the surface region. Especially, it cannot vanish at the resonance frequency ω_r for excitation of a standing plasma wave in the low density surface region in order to produce the singularity of $d_\perp(\omega)$. On the other hand, near ω_r the induced charge density will exhibit a spatial oscillation in the surface region /3.31/ and the numerator of (5.44), measuring the dipole moment of this charge distribution, will diverge at resonance, $\omega = \omega_r$, since then the amplitude of this spatial oscillation is enhanced by a factor ($\sim p_t p_\ell/k_x^2$ as discussed at the end of Sect. 5.5) which diverges in the LWL. Moreover, as the frequency sweeps through the resonance, the phase of the excited plasma wave in the surface layer will change, so that the induced dipole moment changes sign at ω_r, whereas the total induced charge is completely insensitive to these surface effects. Thus, we see that the resonant excitation of standing plasma waves in the low density surface region indeed leads, in the LWL, to a pole structure of $d_\perp(\omega)$. Near the pole, $d_\perp(\omega)$ should be interpreted as dipole moment rather than as mean position of the induced charge distribution.

So far our discussion of MSP modes has neglected damping effects. If damping is included, the pole structure of $d_\perp(\omega)$ is smeared out, its imaginary part becomes a broadened δ function peaked at ω_r and its real part exhibits the S-like shape of a smeared out principle value function. The reflection amplitude r_p, (5.93), will no longer diverge for real values of ω and k_x, and surface plasmons are damped. Nevertheless, for sufficiently small damping, a damped eigenmode is expected to lead still to an enhanced response of the system, so that $|r_p|^2$ should exhibit a more or less pronounced maximum instead of the singularity which in the absence of damping defines the dispersion of the eigenmode.

5.6.3 Experimental Evidence for Multipole Surface Plasmons

"Multipole" surface plasmons are known to exist at coated metal surfaces such as aluminum with an overlayer of adsorbed alkali atoms /3.84 ; 5.12 - 15/, and plasma waves have also been observed in thick adlayers of Ag on an Au substrate /3.35/. For clean metal surfaces, on the other hand, neither experiments nor FEIBELMAN's RPA calculations /3.84/ seemed to indicate the existence of such modes /3.84 ; 5.13/.

But recently it became clear that the photoyield experiments on aluminum /3.90 , 94/, which are in excellent agreement with FEIBELMAN's RPA theory /3.84 , 90/, provide indirect evidence for the importance of an additional surface plasmon mode for the optical response properties of a clean aluminum surface. As discussed in Sect. 3.10, KEMPA and FORSTMANN /3.33 ; 5.16/ explained (with a hydrodynamic model calculation) the experimental photoyield spectrum by a local plasmon excitation in the surface region of reduced electron density. SCHWARTZ and SCHAICH /3.32/ showed numerically that this explanation requires model parameters which insure the existence of a MSP mode. Finally, KEMPA and GERHARDTS /3.34/ pointed out that the frequency dependence of Feibelman's $d_\perp(\omega)$ for each r_s value he considered reveals a broadened pole structure which can be related to a damped MSP mode. Since for r_s = 2 Feibelman's theory is in excellent agreement with the experimental photoyield spectrum of aluminum, which shows a pronounced maximum in just the frequency region where $d_\perp(\omega)$ exhibits the pole structure, the existence of this surface excitation mode and its importance for the response properties of a clean Al surface are evident.

Since it contributes to an understanding of the frequency dependence of $d_\perp(\omega)$, we recall the discussion of the pole structure near 0.85 ω_p given by KEMPA and GERHARDTS /3.34/. In Fig. 5.1 we reproduce their comparison of Feibelman's results based on the self-consistent Lang - Kohn density profile for r_s = 4 (curves "1" of Fig. 5.1) with hydrodynamic calculations for a double-step model of the electron density simulating the diffuseness of the surface. The width of the surface layer (d = 1.5 Å), its plasma frequency (ω_{ps} = 0.5 ω_p), and the damping in the bulk (γ = 0) were so chosen that the broadened pole structure of the hydrodynamic model calculations (curves "2" and "3" of Fig. 5.1a) appears somewhere slightly above ω = 0.8 ω_p The peaks of the absorptance curves shown in Fig. 5.1a are completely determined by the respective peaks of Im{$d_\perp(\omega)$}, whereas the prefactor with a broad maximum at ω = 0.71 ω_p has little effect on the peak position. Near the peak of Im{$d_\perp(\omega)$}, the curve for the real part of $d_\perp(\omega)$ shows a broadened principal value structure superimposed on a monotonically increasing background function which becomes large at ω_p.

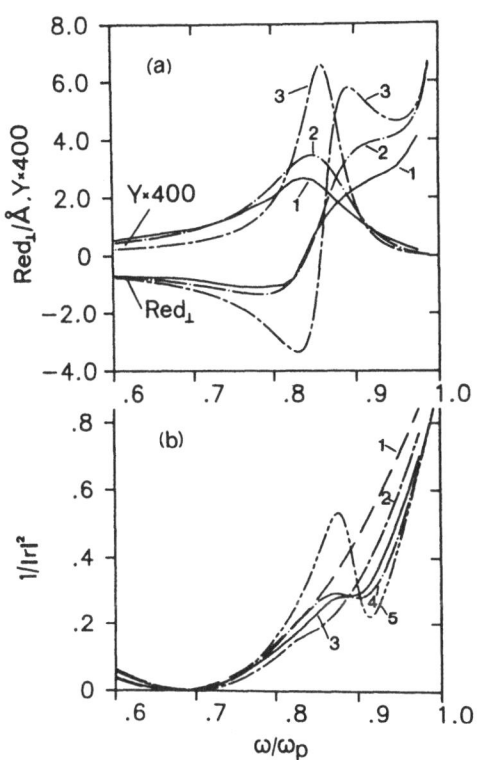

Fig. 5.1. (a) Surface response func-
tions $Re\{d_\perp(\omega)\}$ and $Y(\omega) = \sqrt{8}(\omega/c)\cdot$
$[(\omega/\omega_p)^2 - 1] Im\{d_\perp(\omega)\}$: FEIBELMAN's
/3.84/ RPA result (1) for $r_s = 4$, and
results from the hydrodynamic approximation (with surface step, see text)
for surface damping $\gamma_s = 0.3 \omega_p$ (2),
and $\gamma_s = 0.15 \omega_p$ (3). (b) Inverse re-
flectivity $1/|r|^2$ calculated with $d_\perp$
values from (a) for k_x values corresponding to ATR configuration: Curves
1,2,3 are obtained from Feibelman's $d_\perp$
for $k_x/k_F = 0.005, 0.05, 0.1$, respectively; curves 4 and 5 from the HD for
$k_x/k_F = 0.1$, and $\gamma_s = 0.3 \omega_p$ and $0.15 \omega_p$,
respectively. From /3.34/

The latter is readily understood as the center of gravity of the induced charge.
For low frequencies it has negative values which means a position in the tail region outside the positive background charge, as has already been calculated by LANG
and KOHN /5.17/ for the static response. With increasing frequency the induced
charges penetrate deeper into the metal, i.e., their mean position assumes larger
values and eventually diverges as ω approaches ω_p, if no bulk damping is included
(as in Feibelman's calculation). Bulk damping will limit the penetration of induced charges and, therefore, the value of $Re\{d_\perp(\omega)\}$ near ω_p. KEMPA and GERHARDTS
/3.34/ emphasize that good agreement between the hydrodynamic and Feibelman's RPA
results for $d_\perp(\omega)$ can be achieved only, if the damping γ_s in the surface layer is
taken much larger than the bulk damping ($\gamma_s > 0.3 \omega_p \gg \gamma$). For larger bulk damping
the raise of the $Re\{d_\perp(\omega)\}$ curve near ω_p is not resolved /5.16/. This finding is
consistent with Feibelman's including electron-hole excitations as the only damping
mechanism. Owing to symmetry breaking, optical excitation of electron-hole pairs
is possible (and effective) in the surface region, but not in the bulk.

Fig. 5.1b shows the reflection coefficient in the surface regime ($k_x > \omega/c$).
Curves "1" to "3", calculated from (5.93) with Feibelman's $d_\perp(\omega)$-result for $r_s = 4$,

show that with increasing values of k_x a structure in $1/|r_p|^2$ becomes apparent which is related to the broadened pole structure of $d_\perp(\omega)$. From similar trends in the hydrodynamic model calculation and from the evolution of the structure towards a zero for smaller damping (curves "4" and "5") it is seen that the structure is the remnant of a damped "multipole" surface plasmon /3.34/. The "regular" surface plasmon is seen in Fig. 5.1b as a broad minimum with zero value (it is non-zero for $\gamma > 0$) slightly below $\omega/\omega_p = 0.7$. It is not related to any structure of $d_\perp(\omega)$. For $r_s = 2$ the pole structure of FEIBELMAN's /3.84/ $d_\perp(\omega)$ is a little less pronounced than for $r_s = 4$, but clearly visible. Thus, a strongly damped "multipole" surface plasmon exists for $r_s = 2$ (Al), too.

This result is not really in conflict with the calculation of INGLESFIELD and WIKBORG /5.12/, who used a single-step model for the electron potential at a clean Al surface. The resulting density profile is similar to that for the IBM and much steeper than the self-consistent Lang - Kohn profile. RPA calculations for the IBM /4.30/ also yield no MSP mode for r_s values in the metallic range (2 to 6) /3.34/, since the surface region of low density is too small. This region increases with increasing values of r_s, and for very large r_s values (low bulk densities) MSP modes appear in this quantum mechanical model, too /3.34/. Recent microscopic calculations for charged surfaces support these arguments /4.32/.

Finally we want to emphasize an important difference between hydrodynamic and RPA calculations, namely the role played by damping effects. In hydrodynamic calculations damping (γ, γ_s) appears as a free parameter, which can be neglected completely $(\gamma = \gamma_s = 0)$, as in the early discussions of MSP. In RPA calculations one damping mechanism is included automatically, namely optical excitation of electron-hole pairs in the surface region, owing to the breaking of translational invariance. The strength of this damping effect depends sensitively on the shape of the electron density profile /4.32/. For the Lang - Kohn profile, and probably for real metals, the damping effects are so large that the MSP modes may easily be overlooked.

Owing to the strong damping, direct observation of "multipole" surface plasmons, e.g. in an ATR experiment, may be hard, if not impossible. On the other hand, it should be possible to excite these modes, in contrast to the "regular" surface plasmon, even at a perfectly flat surface by incident light, because they are related to a peak in $\text{Im}\{d_\perp(\omega)\}$, i.e., in the absorptance. Therefore, excitation of a MSP should lead to a reduced intensity of reflected light. As pointed out recently /4.32/ the MSP resonance, i.e. position and height of the maximum of absorptance and photo-yield spectrum, should be tunable by an applied static electric field which modifies the electronic surface profile (cf. Fig. 4.1).

5.7 Résumé

The derivation of the surface response functions $d_\perp(\omega)$ and $d_\parallel(\omega)$ was based on the fact that far away from the surface only the classical transverse electromagnetic fields survive. The idea was now to replace the standard boundary conditions of classical Fresnel optics by new generalized boundary conditions for the transverse fields, which take the nonlocal effects in the surface region into account.

Summarizing the discussion of Sects. 5.2, 3, we can write these generalized boundary conditions in the long-wavelength limit as

$$D_z^>(0) - D_z^<(0) = i(\varepsilon_t - \varepsilon_a)k_x d_\parallel(\omega) E_x^>(0) \quad , \tag{5.102a}$$

$$E_x^>(0) - E_x^<(0) = i(\frac{1}{\varepsilon_a} - \frac{1}{\varepsilon_t})k_x d_\perp(\omega) D_z^>(0) \quad , \tag{5.102b}$$

for p-polarized light and as

$$B_x^>(0) - B_x^<(0) = i(\varepsilon_t - \varepsilon_a)\frac{\omega}{c} d_\parallel(\omega) E_y^>(0) \quad , \tag{5.103a}$$

$$E_y^>(0) - E_y^<(0) = 0 \quad , \tag{5.103b}$$

for s-polarized light. From a macroscopic point of view, the surface response functions $d_\perp(\omega)$ and $d_\parallel(\omega)$ define new boundary conditions from which the exact asymptotic fields, including nonlocal surface effects, can be calculated for arbitrary direction of the incident light. But it should be clear that (5.102, 103) provide only a frame-work for the discussion, not a solution of the problem of nonlocal optics, since $d_\perp(\omega)$ and $d_\parallel(\omega)$ are not known a priori. In order to calculate these quantities, one has to go back to a model of the optical surface response. This may be a microscopic model, such as RPA response of a jellium surface, or a phenomenological model as, for instance, the hydrodynamic approximation with a stepped-density model and additional boundary conditions. In Sects. 5.4-6 we discussed such model calculations and tried to get a physical understanding of the frequency dependence of $d_\perp(\omega)$.

The conceptual difference between the arguments leading to (5.102, 103) and the hydrodynamic approximation discussed in Chap. 2 should be emphasized. In the HD longitudinal electric fields related to plasma waves are considered in addition to the classical transverse electromagnetic fields, the standard boundary conditions are assumed to remain valid for the total fields and additional boundary conditions are formulated. The boundary conditions of the HD are simple and explicit and, in contrast to (5.102, 103), they are independent of frequency and angle of incidence, and they allow to calculate the fields everywhere in space and for all frequencies, without restriction to the LWL.

The LWL generalized boundary conditions (5.102, 103), on the other hand, involve only the asymptotic transverse fields and give no information about the actual fields in the surface region. But they relate, without any specific model assumptions, observable quantities such as reflectivity or surface-plasmon dispersion relation to the surface-response functions describing nonlocal effects. Therefore, it is useful to evaluate experimental results on metal surfaces, to which the LWL applies, in terms of $d_\perp(\omega)$ and $d_\parallel(\omega)$. Apart from being model independent, such a representation of experimental results is economical, since irrelevant angular dependencies are avoided, and since many different observable quantities are determined by the same surface response functions.

Theoretical calculations of the surface response functions $d_\perp(\omega)$ and $d_\parallel(\omega)$ have been performed within the phenomenological hydrodynamic approximation /3.34 , 52 ; 5.16/ and also within the microscopic random-phase approximation /3.84 ; 4.30 , 32/. Recently it has been emphasized /5.18 - 21/ that the use of the RPA together with a self-consistent Lang - Kohn - type effective potential is not quite consistent, since exchange and correlation effects for the responding system are not treated on the same footing as in the ground state. Improved calculations within a "time dependent local density approximation" have been performed for the low-frequency longitudinal response /5.20 , 21/, and very recently for the high-frequency regime of interest in metal optics near the plasma frequency /5.22/. These recent calculations confirm qualitatively the corresponding RPA results /4.32 ; 5.22/.

References

Chapter 1

1.1   M. Born, E. Wolf: *Principles of Optics* (Pergamon, Oxford 1975)
1.2   P.J. Feibelman: Progr. Surface Sci. **12**, 287 (1982)
1.3   D. Pines: *Elementary Excitations in Solids* (Benjamin, New York 1964)
1.4   H. Raether: "Excitation of Plasmons and Interband Transitions by Electrons", in Springer Tracts in Modern Physics, Vol. 88 (Springer, Berlin, Heidelberg, New York 1980)
1.5   F. Sauter: in "Fachberichte der Physikertagung der DPG", Düsseldorf 1964
1.6   F. Sauter: Z. Phys. **203**, 488 (1967)
1.7   F. Forstmann: Z. Phys. **203**, 495 (1967)
1.8   A.R. Melnyk, M.J. Harrison: Phys. Rev. Lett. **21**, 85 (1968)
1.9   I. Lindau, P.O. Nilsson: Phys. Lett. **A31**, 352 (1970)
1.10 M. Anderegg, B. Feuerbacher, B. Fitton: Phys. Rev. Lett. **27**, 1565 (1971)
1.11 P. Apell, Å. Ljungbert, S. Lundqvist: Physica Scripta **30**, 367 (1984)

Chapter 2

2.1   F. Sauter: in "Fachberichte der Physikertagung der DPG", Düsseldorf 1964
2.2   F. Sauter: Z. Phys. **203**, 488 (1967)
2.3   G.D. Mahan: *Many Particle Physics* (Plenum, New York 1981)
      D. Pines, P. Nozières (eds.): *The Theory of Quantum Liquids* (Benjamin, New York 1966)
2.4   V.M. Agranovich, V.L. Ginzburg: *Spatial dispersion in crystal optics and the theory of excitons* (Wiley, New York 1966)
2.5   K. Sturm: Adv. in Physics **31**, 1 (1982)
2.6   F. Forstmann, H. Stenschke: Phys. Rev. Lett. **38**, 1365 (1977)
2.7   M. Born, E. Wolf: *Principles of Optics* (Pergamon, Oxford 1975)
2.8   J.D. Jackson: *Classical Electrodynamics*, 2nd ed. (Wiley, New York 1975)
2.9   V.M. Agranovich, D.L. Mills (eds.): *Surface Polaritons* (North Holland, Amsterdam 1982)
2.10 F. Forstmann: Z. Phys. B **32**, 385 (1979)
2.11 F. Bloch: Z. Phys. **81**, 363 (1933)
2.12 D. Pines, P. Nozières (eds.): *Theory of Quantum Liquids* (Benjamin, New York 1966)
2.13 R. Becker, F. Sauter: *Theorie der Elektrizität*, Vol. III (Teubner, Stuttgart 1969) § 64
2.14 F. Bloch: Helv. Physica Acta **1**, 385 (1934)
2.15 Ref. 2.8 §§ 10.8, 9
2.16 D. Pines: *Elementary Excitations in Solids* (Benjamin, New York 1964)
2.17 G. Barton: Rep. Prog. Phys. **42**, 963 (1979)
2.18 S.I. Pekar: Zh. Eksp. Teor. Fiz. **33**, 1022 (1957) [Sov. Phys. JETP **6**, 785 (1958)], ibid. **34**, 1176 (1958) [ibid **7**, 813 (1958)]
2.19 G.S. Agarwal, D.N. Pattanayak, E. Wolf: Phys. Rev. **B 10**, 1447 (1974)

2.20 J.L. Birman, J.J. Sein: Phys. Rev. **B 6**, 2482 (1972)
2.21 A.A. Maradudin, D.L. Mills: Phys. Rev. **B 7**, 2787 (1973)
2.22 M.F. Bishop, A.A. Maradudin: Phys. Rev. **B 14**, 3384 (1976)
2.23 M. Grossmann, J. Biellmann, S. Nikitin: "Test of Validity of Spatial Disper-
      sion Theories on Lead Jodide Crystal Spectra", in Springer Tracts in Modern
      Physics, Vol. 73 (Springer, Berlin, Heidelberg, New York 1975) p. 242-264
2.24 A.D. Boardman, R. Ruppin: Surface Sci. **112**, 153 (1981)
2.25 K.L. Kliewer: Phys. Rev. **B 14**, 1412 (1976)
2.26 L.D. Landau, E.M. Lifschitz: *Lehrbuch der Theor. Physik*, Band VI (Akademie
      Verlag, Berlin 1974) § 64
2.27 G. Mukhopadhyay, S. Lundqvist: Solid State Commun. **25**, 881 (1978)
2.28 F. Forstmann, H. Stenschke: Phys. Rev. **B 17**, 1489 (1978)
2.29 R. Kötz, D.M. Kolb, F. Forstmann: Surface Sci. **91**, 489 (1980)
2.30 F. Abelès, Y. Borensztein, M. De Crescenzi, T. Lopez-Rios: Surface Sci. **101**,
      123 (1980)
2.31 P.B. Johnson, R.W. Christy: Phys. Rev. **B 6**, 4370 (1972)

Chapter 3

3.1   F. Forstmann: Z. Physik **203**, 495 (1967)
3.2   F. Sauter, F. Forstmann, K. Sturm: Helv. Phys. Acta **41**, 1138 (1968)
3.3   J. Lagois, B. Fischer: "Introduction to Surface Exciton Polaritons", in
      Festkörperprobleme - Advances in Solid State Physics, Vol. **18**, J. Treusch (ed.)
      (Vieweg, Braunschweig 1978) p. 197
3.4   A. Stahl, C. Uihlein: "Optical Boundary Value Problem in Spatially Dispersive
      Media", in Festkörperprobleme - Advances in Solid State Physics, Vol. **19**,
      J. Treusch (ed.) (Vieweg, Braunschweig 1979) p. 159
3.5   A.R. Melnyk, K.M.J. Harrison: Phys. Rev. Lett. **21**, 85 (1968); Phys. Rev. **B 2**,
      835 (1970)
3.6   I. Lindau, P.O. Nilsson: Phys. Lett. **A 31**, 352 (1970); Physica Scripta **3**,
      87 (1971)
3.7   M. Anderegg, B. Feuerbacher, B. Fitton: Phys. Rev. Lett. **27**, 1565 (1971)
3.8   P.J. Feibelman: Phys. Rev. Lett. **35**, 617 (1975)
3.9   R.H. Ritchie: Phys. Rev. **106**, 874 (1957); Progr. Theor. Phys. **29**, 607 (1963)
3.10 R.H. Ritchie: Surface Sci. **34**, 1 (1973)
3.11 H. Raether: Physics of Thin films **9**, 147 (1977)
3.12 A.A. Maradudin: "Surface Waves", in Festkörperprobleme - Advances in Solid
      State Physics, Vol. **21**, J. Treusch (ed.) (Vieweg, Braunschweig 1981) p. 425
3.13 A.D. Boardman (ed.): *Electromagnetic Surface Modes* (Wiley, New York 1982)
3.14 J. Zenneck: Ann. Physik **28**, 665 (1909);
      A. Sommerfeld: *Vorlesungen über Theor. Physik*, Band VI (Akademische Verlags-
      ges., Leipzig 1966) § 32
3.16 E.A. Stern: Rand Report P 2270
3.17 A. Otto: Z. Phys. **185**, 232 (1965)
3.18 F. Abelès: in *Advanced Optical Techniques*, ed. by A.C.S. von Heel (North
      Holland, Amsterdam 1967)
3.19 A. Otto: in *Optical properties of Solids. New Developments*, ed. by B.O.
      Seraphin (North Holland, Amsterdam 1976) p. 678
3.20 K. Sturm: Z. Phys. **209**, 329 (1968)
3.21 J. Harris, A. Griffin: Phys. Lett. **37A**, 387 (1971);
      J. Harris: Phys. Rev. **B 4**, 1022 (1971)
3.22 K.J. Krane, H. Raether: Phys. Rev. Lett. **37**, 1355 (1976)
3.23 A.J. Bennett: Phys. Rev. **B 1**, 203 (1970)
3.24 A. Tadjeddine, D.M. Kolb, R. Kötz: Surface Sci. **101**, 277 (1980)
3.25 A. Eguiluz, S.C. Ying, J.J. Quinn: Phys. Rev. **B 11**, 2118 (1975)
3.26 A. Eguiluz, J.J. Quinn: Phys. Lett. **A 53**, 151 (1975)
3.27 A. Eguiluz, J.J. Quinn: Phys. Rev. **B 14**, 1347 (1976)
3.28 J.J. Quinn, A. Eguiluz, R.F. Wallis, S. Das Sarma: J. Vac. Sci. Technol. **19**,
      402 (1981)

3.29 A.D. Boardman, B.V. Paranjape, R. Teshima: Phys. Lett. **48A**, 327 (1974)
3.30 A.D. Boardman, B.V. Paranjape, R. Teshima: Surface Sci. **49**, 275 (1975)
3.31 A.D. Boardman, B.V. Paranjape, Y.O. Nakamura: Phys. Stat. Sol. (b) **75**, 347 (1976)
3.32 C. Schwartz, W.L. Schaich: Phys. Rev. **B 30**, 1059 (1984)
3.33 K. Kempa, F. Forstmann: Surface Sci. **129**, 516 (1983)
3.34 K. Kempa, R.R. Gerhardts: Solid State Commun. **53**, 579 (1985)
3.35 G. Piazza, D.M. Kolb, K. Kempa, F. Forstmann: Solid State Commun. **51**, 905 (1984)
3.36 P.J. Feibelman: Phys. Rev. **176**, 551 (1968); Phys. Rev. **B 3**, 220 (1971)
3.37 T. Lopez-Rios, F. Abelès, G. Vuye: J. de Physique **40**, L 343 (1979)
3.38 T. Lopez-Rios, M. De Crescenzi, Y. Borensztein: Solid State Commun. **30**, 755 (1979)
3.39 M. De Crescenzi, T. Lopez-Rios, G. Vuye, N.J. Mansov, Y. Borensztein: Thin Solid Films **57**, 89 (1979)
3.40 D.M. Kolb: J. de Physique **38**, C 5 (1977)
3.41 J.D.E. McIntyre: in *Advances in Electrochemistry and Electrochem. Engineering*, ed. by R.H. Muller, Vol. 9 (Wiley, New York 1973) p. 61
     J.D.E. McIntyre: Surface Sci. **37**, 658 (1973)
3.42 F. Forstmann, K. Kempa, D.M. Kolb: J. Electroanal. Chem. **150**, 241 (1983)
3.43 F. Forstmann, R. Kötz, D.M. Kolb: in Proceedings of the 8. Intern. Vacuum Congress, F. Abelès, M. Croset (eds.) (Soc. Francaise du Vide, Cannes, France 1980) p. 425
3.44 D.A. Aspnes: in *Optical Properties of Solids. New Developments*, ed. by B.O. Seraphin (North Holland, Amsterdam 1976) p. 799
3.45 R.H. Müller: Advances in Electrochem. and Electrochem. Engin. **9**, 164 (1973)
3.46 B.E. Hayden, E. Schweizer, R. Kötz, A.M. Bradshaw: Surface Sci. **111**, 26 (1981)
3.47 F. Abelès, T. Lopez-Rios: Surface Sci. **96**, 32 (1980)
3.48 K. Kempa, F. Forstmann, R. Kötz, B.E. Hayden: Surface Sci. **118**, 649 (1982)
3.49 R.C. O Handley, D.K. Burge: Surface Sci. **48**, 214 (1975)
3.50 F. Chao, M. Costa: Surface Sci. **135**, 497 (1983); **157**, L 328 (1985)
3.51 K. Kempa: Surface Sci. **157**, L 323 (1985)
3.52 K. Kempa, R.R. Gerhardts: Surface Sci. **150**, 157 (1985)
3.53 G. Mie: Ann Phys. (Leipzig) **25**, 377 (1908)
3.54 U. Kreibig: J. Phys. **F 4**, 999 (1974)
3.55 L. Genzel, T.P. Martin, U. Kreibig: Z. Phys. **B 21**, 339 (1975)
3.56 U. Kreibig, A. Althoff, H. Presmann: Surface Sci. **106**, 306 (1981)
3.57 M.A. Smithard: Solid State Commun. **13**, 153 (1973)
3.58 J.D. Ganiêre, R. Rechsteiner, M.A. Smithard: Solid State Commun. **16**, 113 (1975)
3.59 H. Abe, W. Schulze, B. Tesche: Chem. Physics **47**, 95 (1980)
3.60 A. Schmitt-Ott, P. Schurtenberger, H.C. Siegmann: Phys. Rev. Lett. **45**, 1284 (1980)
3.61 J.A.A.J. Perenboom, P. Wyder, F. Meier: Physics Reports **78**, 173 (1981)
3.62 J.C. Maxwell-Garnett: Phil. Trans. Roy. Soc. **203**, 385 (1904); **205**, 237 (1906)
3.63 C.G. Granqvist, O. Hunderi: Phys. Rev. **B 16**, 3513 (1977), **B 18**, 2897 (1978); Z. Phys. **B 30**, 47 (1978)
3.64 A. Liebsch, B.N.J. Persson: J. Phys. C: Solid State Phys. **16**, 5375 (1983)
3.65 J.C. Garland, D.B. Tanner (eds.): *Electrical Transport and Optical Properties of Inhomogeneous Media* (AIP, New York 1978)
3.66 R. Burridge, S. Childress, E. Topanicolaou (eds.): *Macroscopic Properties of Disordered Media*, Lecture Notes in Physics, Vol. 154 (Springer, Berlin, Heidelberg, New York 1982)
3.67 e.g. R. Becker, F. Sauter: *Theorie der Elektrizität*, Vol. I (Teubner, Stuttgart 1957)
3.68 R. Ruppin: in Ref. /3.13/ p. 345
3.69 A. Boardman, B.V. Paranjape: J. Phys. **7**, 1935 (1977)
3.70 R. Ruppin: Phys. Rev. **B 11**, 2871 (1975)
3.71 A.A. Lushnikov, A.J. Simonov: Phys. Lett. **44A**, 45 (1973); Z. Phys. **270**, 17 (1974)

3.72 A.A. Lushnikov, V.V. Maksimenko, A.J. Simonov: in *Electromagnetic Surface Modes*, ed. by A.D. Boardman (Wiley, New York 1982) p. 305
3.73 R. Ruppin, Y. Yatom: Phys. Stat. Sol. (b) **74**, 647 (1976)
3.74 W. Ekardt, D.B. Trans Thoai, F. Frank, W. Schulze: Solid State Commun. **46**, 571 (1983)
3.75 P. Ascarelli, M. Cini: Solid State Commun. **18**, 385 (1976)
3.76 P. Apell, A. Ljungbert: Solid State Commun. **44**, 1367 (1982); Physica Scripta **26**, 113 (1982)
3.77 G. Mukhopadhyay, S. Lundqvist: Solid State Commun. **44**, 1379 (1982)
3.78 The essential results of Ref. /3.69/ are not changed, when the boundary conditions discussed in Chap. 2 are employed
3.79 R. Ruppin: Solid State Commun. **20**, 17 (1976); J. Opt. Soc. Am. **66**, 449 (1976)
3.80 A. Kawabata, R. Kubo: J. Phys. Soc. Japan **21**, 1765 (1966)
3.81 C. Kittel: *Quantum Theory of Solids* (Wiley, New York 1963) p. 48
3.82 F. Fujimoto, K. Komaki: J. Phys. Soc. Japan **25**, 1679 (1968)
3.83 K.L. Kliewer: Surface Sci. **101**, 57 (1980)
3.84 P.J. Feibelman: Progr. Surface Sci. **12**, 287 (1982)
3.85 P.J. Feibelman: Phys. Rev. **B 12**, 1319 (1975)
3.86 P.J. Feibelman: Phys. Rev. **B 23**, 2629 (1981)
3.87 K. Kempa, F. Forstmann: Surface Sci. **129**, 516 (1983);
      F. Forstmann, R.R. Gerhardts: "Metal Optics near the Plasma Frequency", in Festkörperprobleme - Advances in Solid State Physics, Vol. **22**, P. Grosse (ed.) (Vieweg, Braunschweig 1982) p. 291
3.88 P. Apell: Physica Scripta **25**, 57 (1982)
3.89 M.P. Seah, W.A. Dench: Surf. and Interf. Analysis **1**, 1 (1979)
3.90 H.J. Levinson, E.W. Plummer, P.J. Feibelman: Phys. Rev. Lett. **43**, 952 (1979)
3.91 R.E.B. Makinson: Proc. Roy. Soc. **A165**, 367 (1937)
3.92 J.G. Endriz: Phys. Rev. **B 7**, 3464 (1973)
3.93 N. Barberan, J.E. Inglesfield: J. Phys. **C14**, 3114 (1981)
3.94 H. Petersen, S.B.M. Hagström: Phys. Rev. Lett. **41**, 1314 (1978);
      H. Petersen: Z. Phys. **B 31**, 171 (1978)
3.95 N.D. Lang, W. Kohn: Phys. Rev. **B 1**, 171 (1978)
3.96 C. Schwartz, W.L. Schaich: Phys. Rev. **B 26**, 7008 (1982)
3.97 C. Schwartz, W.L. Schaich: J. Phys. **C17**, 537 (1984)
3.98 W. Ekardt: Phys. Rev. **B 32**, 1961 (1985)

Chapter 4

4.1   G.D. Mahan, J.J. Hopfield: Phys. Rev. **A135**, 428 (1964); see also /2.4/ and /2.23/
4.2   G.S. Agarwal, D.N. Pattanayak, E. Wolf: Phys. Rev. B **11**, 1342 (1975) and references therein;
      J.T. Foley, A.J. Devaney: Phys. Rev. B **12**, 3104 (1975)
4.3   D.L. Johnson, P.R. Rimbey: Phys. Rev. B **14**, 2398 (1976)
4.4   K.L. Kliewer, R. Fuchs: Phys. Rev. **172**, 607 (1968);
      R. Fuchs, K.L. Kliewer: Phys. Rev. **185**, 905 (1969)
4.5   K.L. Kliewer, R. Fuchs: Advances in Chemical Physics, ed. by I. Prigogine and S.A. Rice, Vol. 27 (Wiley, New York 1974)
4.6   P.R. Rimbey, G.D. Mahan: Solid State Commun. **15**, 35 (1974)
4.7   F. Flores, F. Garcia-Moliner, R. Monreal: Phys. Rev. B **15**, 5087 (1977)
4.8   F. Garcia-Moliner, F. Flores: J. Physique **38**, 851 (1977)
4.9   P. Ahlqvist, R. Monreal, F. Flores, F. Garcia-Moliner: Physica Scripta **26**, 35 (1982)
4.10 F. Forstmann: Z. Physik **32**, 385 (1979)
4.11 A.K. Rajagopal, F. Forstmann: unpublished
4.12 G.E.H. Reuter, E.H. Sondheimer: Proc. Roy. Soc. Lond. **A195**, 336 (1948)
4.13 K.L. Kliewer: in *Photoemission and the Electronic Properties of Surfaces*, ed. by B. Feuerbacher, B. Fitton, R.F. Willis (Wiley, New York 1978) p. 45
4.14 G. Mukhopadhyay, S. Lundqvist: Physica Scripta **17**, 69 (1978); Solid State Commun. **21**, 629 (1977)

4.15 P. Apell: Physica Scripta **17**, 535 (1978); **23**, 284 (1981)
4.16 R.R. Gerhardts: Physica Scripta **28**, 235 (1983)
4.17 P.R. Rimbey, G.D. Mahan: Solid State Commun. **15**, 35 (1974)
4.18 R. Zeyher, J.L. Birman, W. Brenig: Phys. Rev. B **6**, 4613 (1972)
4.19 W. Hanke: Adv. Phys. **27**, 287 (1978)
4.20 A. Bagchi: Phys. Rev. B **15**, 3060 (1977)
4.21 P.J. Feibelman: Phys. Rev. B **12**, 4282 (1975)
4.22 P.J. Feibelmann: Phys. Rev. B **22**, 3654 (1980)
4.23 N.D. Lang, W. Kohn: Phys. Rev. B **1**, 4555 (1970)
4.24 P. Hohenberg, W. Kohn: Phys. Rev. **136**, B 864 (1964);
      W. Kohn, L.J. Sham: Phys. Rev. **140**, A 1133 (1965)
4.25 T. Maniv, H. Metiu: Phys. Rev. B **22**, 4731 (1980); J. Chem. Phys. **76**, 2697 (1982)
4.26 T. Maniv, H. Metiu: J. Chem. Phys. **76**, 696 (1982)
4.27 G.E. Korzeniewski, T. Maniv, H. Metiu: J. Chem. Phys. **76**, 1564 (1982)
4.28 T. Maniv: Phys. Rev. B **26**, 2856 (1982)
4.29 D.M. Newns: Phys. Rev. B **1**, 3304 (1970)
4.30 R.R. Gerhardts, K. Kempa: Phys. Rev. B **30**, 5704 (1984)
4.31 R.R. Gerhardts, K. Kempa: unpublished
4.32 P. Gies, R.R. Gerhardts: Proc. 2nd Intern. Conf. on Surface Waves, Ohrid, Yugoslavia 1985; Europhys. Lett. **1**, 513 (1986)
4.33 P. Gies, R.R. Gerhardts: Phys. Rev. B **31**, 6843 (1985); B **33**, 982 (1986)
4.34 P. Gies, R.R. Gerhardts, T. Maniv: to be published
4.35 J. Lindhard: Kgl. Dansk Videnskab. Selskab, Mat. Fys. Medd. **28**, No. 8 (1954)
4.36 B.N.J. Persson: J. Phys. C **13**, 435 (1980)
4.37 V. Cataudella, V. Marigliano Ramaglia, G.P. Zucchelli: Phys. Lett. **92A**, 359 (1982)
4.38 G. Mukhopadhyay: Solid State Commun. **28**, 277 (1978)

Chapter 5

5.1   P.J. Feibelmann: Phys. Rev. B **14**, 762 (1976)
5.2   P. Apell: Physica Scripta **24**, 795 (1981)
5.3   A.C.R. van Alkemade: Wied. Ann. **20**, 22 (1883)
5.4   P. Drude: Wied. Ann. **36**, 532 (1889); **43**, 126 (1891)
5.5   H. Mayer: *Physik dünner Schichten*, Bd. 1 (Wiss. Verlagsges. Stuttgart, 1950)
5.6   W.J. Plieth, K. Naegele: Surf. Sci. **64**, 484 (1977)
5.7   A. Bagchi, R.G. Barrera, A.K. Rajagopal: Phys. Rev. B **20**, 4824 (1979)
5.8   J.D.E. McIntyre, D.E. Aspnes: Surf. Sci. **24**, 417 (1971)
5.9   F. Abelès, T. Lopez-Rios: Surf. Sci. **96**, 32 (1980)
5.10 J.E. Sipe: Phys. Rev. B **22**, 1589 (1980)
5.11 A. Eguiluz, J.J. Quinn: Nuovo Cimento **39 B**, 828 (1977)
5.12 J.E. Inglesfield, E. Wikborg: J. Phys. F **5**, 1706 (1975)
5.13 P. Ahlqvist, P. Apell: Physica Scripta **25**, 587 (1982)
5.14 A.U. MacRae, K. Müller, J.J. Lander, J. Morrison, J.C. Phillips: Phys. Rev. Lett. **22**, 1048 (1969)
5.15 S. Andersson, U. Jostell: Surf. Sci. **46**, 625 (1974)
5.16 F. Forstmann, K. Kempa: J. Physique **45** C5, 191 (1984)
5.17 N.D. Lang, W. Kohn: Phys. Rev. B **3**, 1215 (1971); Phys. Rev. B **7**, 3541 (1973)
5.18 B.N.J. Persson, E. Zaremba: Phys. Rev. B **30**, 5669 (1984)
5.19 A. Liebsch: Phys. Rev. Lett. **54**, 67 (1985)
5.20 A.G. Eguiluz: Phys. Rev. B **31**, 3303 (1985)
5.21 A. Liebsch: Phys. Rev. B **32**, 6255 (1985)
5.22 P. Gies, R.R. Gerhardts: Proc. 6th Int. Conf. Solid Surfaces (ICSS-6), Oct. 27 - 31 1986, Baltimore, USA, to be published in J. Vac. Sci. Tech.

Subject Index

Absorption density 16, 55, 87

Additional boundary conditions (ABC)
11, 14, 18, 20, 58, 62, 64, 65, 67,
69, 107

Additional surface forces 63

Adsorbed layer 47

Analytical separation of response
modes 80, 83

Boundary conditions 63, 66, 67, 72,
94, 104

- generalized b.c. for asymptotic
fields 90, 91, 95, 123

Charged oscillator model 65

Colloidal gold 48

Compound system 43, 46

Conductivity 9, 64, 65, 74

Coupling to substrate 43, 46

Dielectric approximation 63, 69

Dielectric constant 10, 19, 41, 49

Dipole length of induced charge density 78, 102, 119

Discontinuity of energy current 63,
69

Drude conductivity 67

- dielectric function 72

Electron-hole excitations 75, 79,
87, 121

Electronic surface profile 75, 77,
78, 87, 118, 122

- of charged surfaces 78, 122

Electroreflectance 37, 43

Eigenmodes

- in a metal layer 28

- in surface layers 36, 40

- on a metal surface 29, 78, 87

Ellipsometry 46, 100

Energy theorem 15

Extended hydrodynamic model 64, 65, 72

Fields near surface 52, 84, 86

Fresnel's formulae 23, 94

Friedel oscillations 53, 75, 79, 83, 87

Green's function 9, 64, 65, 66

Hydrodynamic approximation (HD) 8

Hydrodynamic equation 9, 65

- eigensolutions of 65

Hydrodynamic model 9, 64, 72, 80, 85,
107, 110, 122

Infinite barrier model (IBM) 76, 78

Integral representation of surface fields
71, 83

Lindhard dielectric function 72, 77, 80

- analytical properties of 80

Linear response theory 7, 62, 73

Local density approximation

- for exchange and correlation 75, 78

Local density approximation (cont.)
- time depended 124
Local layer model 90, 103, 104, 111
Long-wavelength limit 10, 75, 78, 89, 96, 113

Material equation 7, 9, 62
Material parameters 10, 64
Maxwell's equations 6, 77, 91
Metal films (on metals) 28, 36, 107, 114, 119
Metal model 10, 67, 69, 72, 107
Microscopic theory 57, 62, 73
Mie resonance 49
Mixed Fourier representation 76
Multipole modes 36, 40, 115, 118, 122

Nonlocality 7
Nonlocal conductivity 10, 62, 64, 65
Nonlocal layer model 107, 111

Optical "surface position" 102

Partial waves method 85
Pekar's ABC 64, 67, 68, 69
Phenomenological approaches 62, 64
Photoemission yield 55
Photoyield spectrum 56, 75, 120
Plasma waves 8, 9, 101, 114
- dispersion 10
- energy current 16
- reflection of 17
Plasmon pole approximation 53, 81, 83
p polarization 13

Random phase approximation 74, 122
Reference fields 91
Reflectance 24, 26, 37, 42
Reflection amplitude 23, 26, 93, 95, 99, 110

Reflection coefficient *see* Reflectance
Resonances 43
- in metal spheres 49
- in surface layers 36, 42, 46, 58
- in thin metal films 28, 59, 114
- with eigenmodes 37, 40
Response functions 9, 65
Response to bare external field 75

SCIB model 70, 77, 79, 82, 87
Shift of eigenfrequency 50
Silver 19, 28, 35, 38, 41, 44
Single particle excitations 54
Singular surface densities 13, 17, 60
Small metal spheres 48
Soft surface 34, 47
Spatial dispersion 7, 65, 67, 69, 82, 112
Specular reflection model 63, 64, 67, 68, 70, 72, 77, 87
- generalizations of 73
s polarization 13
Standard optics 7, 13, 23, 24, 30
Stress boundary condition 108
Surface absorptance 78, 83, 85, 87
Surface absorption mechanism 87
Surface model 11, 64
Surface plasmon dispersion 29, 117, 122
- electrostatic approximation 32
- for soft surface 34
- within standard optics 30
Surface response function $d_\perp(\omega)$ 78, 88, 89, 99, 102, 111, 123
Surface solutions 90, 93
Surface step model 25, 34, 38
Susceptibility 63, 68, 69

Transmission amplitude 23, 93, 95, 99

Use of bulk response properties 11, 63, 64